*Cambridge Studies in Biological and Evolutionary Anthropology 30*

# Human Biology of Pastoral Populations

Animal-herding (pastoralism) is a subsistence strategy that is practiced by populations of low-producing ecosystems worldwide. Increasingly, it is vanishing due to land pressure and ecological degradation, particularly in the developing world. While previous books have examined the social, cultural and economic dimensions of the pastoral way of life, until now there has been no systematic examination of the biology and health of pastoral groups. *Human Biology of Pastoral Populations* fills this gap by drawing together our current knowledge of the biology, population structure and ecology of herding populations. It investigates how pastoral populations adapt to limited and variable food availability, the implications of the herding way of life for reproductive patterns, population structure and genetic diversity and the impacts of ongoing social and ecological changes on the health and well-being of these populations. This volume will be of broad interest to scholars in anthropology, human biology, genetics and demography.

WILLIAM R. LEONARD is Professor of Anthropology at Northwestern University, Illinois. He is a biological anthropologist whose research focuses on aspects of physiology, nutrition and health. He has extensive fieldwork experience in Siberia and Latin America examining how human populations adapt to extreme environments.

MICHAEL H. CRAWFORD is Professor of Anthropology at the University of Kansas. His research focuses on anthropological genetics, particularly in populations of the Americas. He has also written *The Origins of Native Americans* (1998; ISBNs 0521 592801 & 0521 004101), and edited *Different Seasons: Biological Aging in Mennonites of Midwestern United States* (2000).

T0269323

*Cambridge Studies in Biological and Evolutionary Anthropology*

*Series Editors*

HUMAN ECOLOGY
C. G. Nicholas Mascie-Taylor, University of Cambridge
Michael A. Little, State University of New York, Binghamton
GENETICS
Kenneth M. Weiss, Pennsylvania State University
HUMAN EVOLUTION
Robert A. Foley, University of Cambridge
Nina G. Jablonski, California Academy of Science
PRIMATOLOGY
Karen B. Strier, University of Wisconsin, Madison

*Consulting Editors*
Emeritus Professor Derek F. Roberts
Emeritus Professor Gabriel W. Lasker

# Human Biology of Pastoral Populations

EDITED BY

## WILLIAM R. LEONARD
*Department of Anthropology*
*Northwestern University, Evanston, Illinois, USA*

## MICHAEL H. CRAWFORD
*Department of Anthropology*
*University of Kansas, Lawrence, Kansas, USA*

CAMBRIDGE
UNIVERSITY PRESS

CAMBRIDGE UNIVERSITY PRESS
Cambridge, New York, Melbourne, Madrid, Cape Town, Singapore, São Paulo

Cambridge University Press
The Edinburgh Building, Cambridge CB2 8RU, UK

Published in the United States of America by Cambridge University Press, New York

www.cambridge.org
Information on this title: www.cambridge.org/9780521780162

First published 2002
This digitally printed version 2008

*A catalogue record for this publication is available from the British Library*

*Library of Congress Cataloguing in Publication data*

Human biology of pastoral populations / edited by William R. Leonard and Michael
H. Crawford.
  p.   cm. – (Cambridge studies in biological and evolutionary anthropology; 30)
Includes bibliographical references and index.
ISBN 0 521 78016 0
1. Pastoral systems.   2. Physical anthropology.   3. Herders – Anthropometry.
4. Herders – Health and hygiene.   5. Indigenous peoples – Ecology.   I. Leonard,
William R. (William Rowe), 1959–   II. Crawford, Michael H., 1939–   III. Series.
GN407.7.H85 2002
599.9′5 – dc21   2001035313

ISBN 978-0-521-78016-2 hardback
ISBN 978-0-521-08163-4 paperback

# Contents

# Contributors

Helen Alinga Akol
Agropastoral Development Programme, Moroto, Uganda

Cynthia M. Beall
Department of Anthropology, Case Western Reserve University, Cleveland,
Ohio 44106, USA

Rosario Calderón
Department of Animal Biology and Genetics, University of the Basque Country,
48080 Bilbao, Spain

Michael H. Crawford
Department of Anthropology, University of Kansas, 622 Fraser Hall, Lawrence,
Kansas 66045, USA

Victoria A. Galloway
Faculty of Medicine, University of Toronto, Toronto, Ontario, Canada

Melvyn C. Goldstein
Department of Anthropology, Case Western Reserve University, Cleveland,
Ohio 44106, USA

Sandra Gray
Department of Anthropology, University of Kansas, 622 Fraser Hall, Lawrence,
Kansas 66045, USA

I. Hershkovitz
Department of Anatomy and Anthropology, Tel Aviv University, Tel Aviv
69978, Israel

Clare Holden
Department of Anthropology, University College London, Gower Street,
London WC1E 6BT, UK

William Irons
Department of Anthropology, Northwestern University, 1810 Hinman Avenue,
Evanston, Illinois 60208, USA

Evgueni Ivakine
Department of Immunology, University of Toronto, Toronto, Ontario, Canada

Marina Kazakovtseva
Institute of Cytology and Genetics, Russian Academy of Sciences, Novosibirsk, Russia

Eugene Kobyliansky
Department of Anatomy and Anthropology, Tel Aviv University, Tel Aviv 69978, Israel

Saara Lehtinen
Department of Clinical Chemistry, Tampere University Hospital, Tampere, Finland

Terho Lehtimäki
Department of Clinical Chemistry, Tampere University Hospital, Tampere, Finland

William R. Leonard
Department of Anthropology, Northwestern University, 1810 Hinman Avenue, Evanston, Illinois 60208, USA

Juhani Leppäluoto
Department of Physiology, University of Oulu, Oulu, Finland

Paul Leslie
Department of Anthropology, University of North Carolina, 301 Alumni Building, Chapel Hill, North Carolina 27599, USA

Michael A. Little
Department of Anthropology, State University of New York, Binghamton, New York 13902, USA

Pauli Luoma
Oulu Regional Institute of Occupational Health, FIN-90220 Oulu, Finland

Ruth Mace
Department of Anthropology, University College London, Gower Street, London WC1E 6BT, UK

Joseph McComb
Department of Anthropology, University of Kansas, 622 Fraser Hall, Lawrence, Kansas 66045, USA

R. John Mitchell
Department of Human Genetics, LaTrobe University, Bundoora, Australia

Mary Jane Mosher
Department of Anthropology, University of Kansas, 622 Fraser Hall, Lawrence, Kansas 66045, USA

Simo Näyhä
Oulu Regional Institute of Occupational Health, FIN-90220 Oulu, Finland

Ludmilla Osipova
Institute of Cytology and Genetics, Russian Academy of Sciences, Novosibirsk, Russia

Renee L. Pennington
Department of Anthropology, University of Utah, Salt Lake City, Utah 84112, USA

Moses S. Schanfield
Analytic Genetic Testing Center, Denver, Colorado 80231, USA

# 1 The biological diversity of herding populations: an introduction

MICHAEL H. CRAWFORD AND WILLIAM R. LEONARD

This volume evolved from a 1997 symposium held at the American Association for the Advancement of Science (AAAS) Meetings in Seattle, Washington. This symposium drew together the leading scholars of nomadic pastoralists from anthropology, demography, genetics and medicine. They focused upon the ecology and population biology of contemporary herding groups from Africa, Asia, Europe, and the Middle East. While previous publications have summarized sociocultural variation in pastoralist groups (see Barfield, 1993), no previous volume has attempted to merge the ecological, demographic, health, and biological facets of the herding existence.

Nomadic pastoralists are of great fascination to the more sedentary westerners, who tend to romanticize the nomads for their free spirit, apparently unencumbered by geographical and political boundaries. Nomads are envied for their perceived freedom, being able to break camp and move on to the next pasture. Bruce Chatwin (a writer and adventurer) in an essay entitled "It's a Nomad *Nomad* World" (Chatwin, 1996) extended this romantic fascination to hypothesize that humans are naturally migratory and that sedentism is the cause of many of the ills of contemporary society. In this introduction, we begin by addressing several key questions about pastoralists: What is nomadic pastoralism, and why does it exist? Why are nomads constantly on the move? What common features do the nomadic societies share? We will then provide an overview of the chapters in this volume, highlighting several central themes addressed throughout the volume.

## What is nomadic pastoralism?

Nomadic refers to movement, while pastoralism is a type of subsistence. Therefore, in general, nomadic pastoralism refers to populations that specialize in animal herding, which requires periodic movement for

1

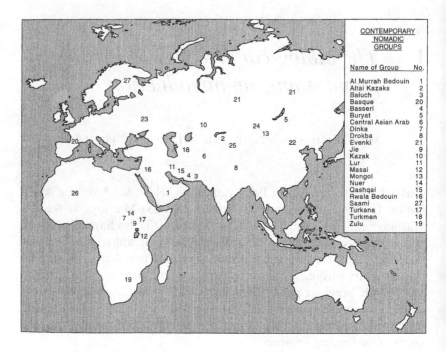

Figure 1.1. Geographical distribution of pastoral populations throughout the Old World.

purposes of grazing. There is enormous variability in herd management strategies, social organization, and degree of mobility. According to Spooner (1973) there are no features of cultural or social organization common to all nomads or even that occur exclusively among nomads.

### Where are the nomads found?

The pastoral nomads are distributed widely throughout the Old World (see Figure 1.1). A large concentration of herders is found in Africa with: the camel and goat herding Berbers of the Sahara (marked number 26 on Figure 1.1), the cattle herding Nilotes, such as the Dinka of the Sudan, and the Masai and Turkana of East Africa. The Middle East contains a number of Bedouin Herders, as shown by the Al Murrah and Rwala Bedouins in Figure 1.1. Europe has only two populations with a history of herding: the Sámi reindeer herders of Finland and the sheep-herding Basques of the Pyrenees of France and Spain. As indicated in Figure 1.1,

Central Asia, because of its extensive grasslands still contains many pastoral populations.

## Why do nomads migrate?

Nomads migrate in order to utilize seasonal pastures more efficiently, avoid hazardous environments, and reduce competition with other groups for resources. Of these three reasons for migration, the dominant motivation is to find fresh pastures for the herds. For example, in the Siberian *taiga*, the Evenki usually winter at a specific location. However, in the early spring, young men are sent out to search for fresh pastures. As soon as these new pastures are located, work parties build corrals and the herds are relocated to the new grazing lands.

In Africa, pastoral groups move away from breeding grounds of disease vectors such as mosquitoes and tsetse flies. In arid regions, where water supplies are at a premium, herders tend to relocate to avoid conflict for pasture and water resources.

## Why does nomadic pastoralism exist?

Pastoralism is an effective means of exploiting marginal environments, such as arid grasslands of the tropics or the tundra and *taiga* ecosystems of the north. In these environments, the amount of energy fixed by plants via photosynthesis (primary productivity) is low (see Begon *et al.*, 1990), and the dominant plants (e.g., grasses, shrubs) are generally poor food sources for humans. The pastoral subsistence economy provides an adaptation to such conditions since it promotes the conversion of low quality plant resources into portable, high quality animal foods. However, the overall low level of energy availability necessitates low population density and high mobility among pastoral populations.

In East Africa there is a continuum from almost total dependence on herds of animals, to some horticulture associated with some cattle herding, to sedentary agriculture. The limiting factors to the type of subsistence practiced by the society are rainfall and soil nutrients. In the case of Arctic/sub-Arctic pastoral groups, such as the Evenki of Siberia, the combination of permafrost and lichen-covered ground does not permit agriculture. Even the hardy onion fails to flourish in Evenkia! Thus, traditionally the Evenki have depended almost entirely on the reindeer for their subsistence. Initially, the Evenki hunted reindeer, later they learned to herd the reindeer, thus, giving them a more reliable source of food, albeit on the hoof! After Russian contact, the Evenki traded furs for Western products, such as tobacco, sugar, and flour. In recent times, antlers have been traded to Japanese merchants, resulting in an "oasis" of Sony television sets, which often get traded to Russian entrepreneurs to obtain vodka.

Today, population growth, environmental degradation and changing patterns of land use threaten many pastoral populations, in part, because it is a subsistence regime that requires large areas of land to support relatively small populations. For example, the creation of new political borders, disrupting the traditional migratory routes can have dire consequences for the survival of the nomadic pastoral way of life. The Kurds of northern Iraq and Turkey have been adversely affected by the creation of fortified political borders thus limiting movements across traditional grazing areas. Environmental pollution in Siberia, associated with oil exploration and massive irradiation has had dire consequences on the reindeer pastorals. Similarly, the political disintegration of the Soviet Union has been paralleled by unprecedented shortages of fuel for helicopters and airplanes, thus, depriving the isolated herders of basic Western staples to which they had grown accustomed.

### What do nomadic pastoralists have in common?

Nomadic pastoralists developed different patterns of social organization that depends on their specific ecological, cultural, political, or historical circumstances. Each geographical region has its own, unique pattern of development and interaction with the surrounding sedentary societies. What all the pastoral nomads share in common is their economic reliance (to varying degrees) on domesticated herds. The periodic migrations of nomadic pastoralists are necessitated by the need for pasture and water for the herds. The distance that the nomads migrate, and the frequency of that migration, are dependent on the ecology of the region. Arid zones require more frequent migrations than the regions where water is plentiful.

Barfield (1993) in his classic volume on pastoral nomads, divided Old World pastoral societies into five distinct zones, each with its own unique style of animal husbandry, ecology, and social organization. In this volume we consider populations from each of Barfield's five zones (South of the Sahara; Desert Zone of camel pastoralism; North of the Arid Desert along the Mediteranean littoral; Eurasian Steppe Zone; High-altitude Pasture of the Tibetan Plateau and neighboring mountain regions). We additionally examine a category of pastoralists not considered by Barfield, the Arctic herders.

## Pastoral zones

### *South of the Sahara*

Cattle herders south of the Sahara in the Sahel and savanna grasslands of East Africa, follow the Great Rift Valley. In this zone cattle are viewed as the most important livestock, yet most of these societies maintain flocks of sheep and goats for subsistence plus donkeys for transport. The pastoralists bordering the northern deserts also include camels in their herds. The herd animals provide blood, milk, and meat for subsistence. Any horticulture practiced in this zone is in the hands of the women. Males traditionally are herders.

In this volume, three of the chapters discuss populations from this pastoral zone: Rene Pennington discusses in Chapter 8 the economic stratification and health among the Herero of Botswana. These Bantu-speaking pastoralists subsist on cattle in the northern Kalahari Desert. Michael Little, in Chapter 7, focuses on the human biology, health, and ecology of Nomadic Turkana pastoralists of northwest Kenya. The Turkana are primarily cattle herders, but also include camels, sheep, goats and donkeys in their herds. In Chapter 5 Sandra Gray and colleagues discuss the adaptive strategies of two East African populations, the Turkana of northwest Kenya and the Karimojong of northeast Uganda. These authors discuss how socioeconomic, political and environmental forces have shaped demographic parameters in these two groups. They view these pastoral adaptive strategies as being reflective on "non-equilibrium" systems; that is, adaptive regimes that are characterized by flexibility and a high degree of seasonal and year-to-year variability.

### *Desert zone of camel pastoralism*

The desert zone of camel pastoralism is contiguous with the Saharan and Arabian Deserts. These pastoralists raise the dromedary camel, on which they rely for both food and transport. These herders trade for dates with the sedentary oasis farmers. In Chapter 4, E. Kobyliansky and I. Hershkovitz focus on the genetic structure (marital patterns, inbreeding, and migration) of the South Sinai Bedouins. Although these Bedouins now make a living in industrial areas of Israel, traditionally, they raised sheep and camels.

### North of the arid desert along the Mediterranean littoral

Pastoralists, north of the arid desert along the Mediterranean littoral, extend through the Anatolian and Iranian Plateaus into the mountains of Central Asia. These nomads take advantage of variation in elevation moving their livestock from the lowland winter pastures to highland summer grazing. Their herds consist of sheep, goats, horses, donkeys, and even camels. Cattle are not usually part of their herds because they require better pasture, more water, and cannot negotiate the mountain trails. In Chapter 11, William Irons, discusses the family organization and demography of the Yomut Turkmen of northern Iran and Afghanistan. These populations are primarily agricultural and herd sheep, goats and camels.

### Eurasian Steppe zone

The Eurasian Steppe pastoral zone includes the horse riding and herding nomads of Central Asia. This zone stretches from the Black Sea to Mongolia, and is largely flat grasslands, punctuated by mountain ranges. Much of this Steppe zone was controlled by nomadic populations a few hundred years ago. Now, agriculture and cattle herding is found together with a complex consisting of horses, sheep, goats, and camels. Crawford and colleagues, Chapter 2, discuss the genetic structure of the Kizhi-Altai population that inhabits this pastoral zone. The Kizhi-Altai are cattle herders, who reside in mountain valleys, covered by grasslands that are conducive to herding. The Kizhi live in yurts that are similar to those of the Mongolian herders.

### High-altitude pasture of the Tibetan Plateau and neighboring mountain regions

This zone is characterized by vast plateaus with grasslands offering rich grazing. The pastoralists on the Tibetan plateau herd yaks, sheep, goats, horses and yak/cattle hybrids. Chapter 6, by Melvyn Goldstein and Cynthia Beall, focuses on the changing patterns of nomadic pastoralism among the Phala nomads of the Tibetan Plateau. The Phala herd yak, sheep, goats, and horses. Like Gray and colleagues, Goldstein and Beall highlight how sociopolitical changes in Tibet have had a profound impact on the life of the Phala nomads.

### Arctic herders

The Nomadic pastoralists of the northern latitudes raise reindeer, an animal that cannot survive outside the tundra/*taiga* zones. However, other domesticated animals cannot survive by grazing on lichens. The arctic zone contains a wide continuum of arctic reindeer exploitation that ranges from hunting to raising and herding. In Chapter 9, Leonard and colleagues examine how historic social and economic changes in Russia have influenced the nutrition and health of the Evenki reindeer herders of central Siberia. Similarly, Crawford and colleagues in Chapter 2, examine the genetic structure of the Evenki and compare it with the patterns observed in the Kizhi-Altai. Chapter 10, by Simo Näyhä and his colleagues, examines the chronic disease patterns observed among the Sámi reindeer herders of Finland.

### Other pastoral groups

This subdivision of marginal ecosystems into these distinct zones excludes various pastoral/agricultural groups in Europe and South America. For example, a case can be made for including the Basque sheep-herders of the Pyrenees. In fact, Rosario Calderón in Chapter 3 examines the genetic structure of the Basque populations of northern Spain. Other pastoral/ agricultural groups in the Alps of Switzerland and Italy exist, but, visualizing Heidi as a pastoral nomad, challenges scientific credulity! However, populations of camelid herders subsisting in South America, should qualify as pastoralists based upon their existence in a marginal, high-altitude environment.

### Alternative perspectives on the biology of pastoral groups

This volume is topically organized into three sections: (1) demography, genetics, and population structure; (2) ecology and health; (3) biocultural and evolutionary perspectives. The chapters by Crawford and colleagues, Calderón, Kobyliansky and Hershkovitz, and Gray and colleagues all demonstrate how the genetic and demographic structure of pastoral populations are shaped by aspects of their distinctive subsistence ecology. Gray and colleagues demonstrate how adapting to "nonequilibrial" ecosystems strongly influences demographic parameters among the Turkana and Karamojong populations. Similarly, Crawford and colleagues show how the clan-based herding systems of indigenous Siberian populations (the Evenki and Kizhi-Altai) help to structure the genetic diversity in these groups.

Chapters 6 through 10 examine aspects of ecology and health among pastoral groups. These chapters consider how pastoral adaptive strategies influence such factors as dietary consumption, growth and development, fertility, mortality and disease risks. Moreover, they demonstrate how changes in traditional pastoral lifeways influence both human and environmental health. Goldstein and Beall demonstrate that traditional pastoral systems of Tibetan nomads successfully maintain their high-altitude grassland ecosystems. In contrast, the "modernization" strategies proposed by the Chinese government threaten to degrade these ecosystems by limiting mobility of the pastoral populations.

Little shows how the biology of traditionally living Turkana pastoralists is highly responsive to environmental variation. Patterns of physical growth, work and fertility are all shaped by seasonal and annual environmental fluctuations.

Pennington explores the influence of economic variation on the growth of Herero children. She finds that the Herero children have growth patterns similar to those observed among other African pastoral groups such as the Turkana. However, variation in economic success, as measured by herd size, was not a significant predictor of childhood growth status.

The chapters by Näyhä and colleagues, and Leonard and colleagues examine how aspects of lifestyle change are influencing health and disease patterns in pastoral populations of northern latitudes. Näyhä and colleagues demonstrate that chronic disease rates among the Sámi of Finland are increasing, but are still considerably lower than those observed in the general population of Finland. Leonard and colleagues find similar results among the Evenki of central Siberia, as lifestyle changes are increasing rates of obesity and risk of coronary heart disease, particularly among women.

The final two chapters by Irons and Holden and Mace explore evolutionary aspects of the pastoral lifeway. Irons shows how, among the Yomut Turkmen, the subordinate social role of women contributes to lower female life expectancy. This study effectively shows how social organization and behavioral factors, in addition to broad ecological constraints, can have a dramatic influence on the demography of pastoral populations.

Holden and Mace examine the relationship between pastoralism and the evolution of adult lactose digestion. They test alternative models for explaining lactase persistence, and demonstrate that simple ecological/geographical models alone are not sufficient to explain the world-wide variation in this parameter. Rather, phylogeny (i.e., shared common ancestry of populations) plays an important role that must be controlled for when studying this trait.

Overall, the chapters in this volume highlight the many avenues through which pastoral subsistence ecology shapes human biology. Adaptation to unstable, nonequilibrial ecosystems contribute to distinctive patterns of growth and development, physical work, fertility and mortality among pastoral populations. Moreover, the high mobility, low population density and familial/clan structure that characterize most pastoral societies have important consequences for their genetic structure and diversity. The diverse adaptive strategies employed by pastoral groups have proved successful in exploiting marginal and often hostile environments; however, throughout the world today, social, economic and political pressures are threatening the persistence of the pastoral lifeway. Further research on the biological dimensions of pastoral subsistence is therefore needed to understand how to best promote the health and well-being of these populations in the face of ongoing change.

### References

Barfield T.J. (1993). *The Nomadic Alternative*. Englewood Cliffs, NJ: Prentice-Hall.
Begon, M., Harper, J.L., and Townsend, C.R. (1990). *Ecology: Individuals, Populations and Communities, 2nd edn.* Boston: Blackwell.
Chatwin, B. (1996). It's a Nomad *Nomad* World. In *Anatomy of Restlessness. Selected Writings 1969–1989*, ed. J. Borm and M. Graves, pp. 100–106, New York: Viking.
Spooner, B. (1973). *The Cultural Ecology of Pastoral Nomads*. Reading, MA: Addison-Wesley.

# 2 Genetic structure of pastoral populations of Siberia: the Evenki of central Siberia and the Kizhi of Gorno Altai

MICHAEL H. CRAWFORD, JOSEPH MCCOMB,
MOSES S. SCHANFIELD AND R. JOHN MITCHELL

The genetic structure of human populations has been defined a number of different ways depending upon the underlying theoretical orientation (Crawford, 1998). Some researchers define genetic structure as the corrections of ideal conditions specified by Hardy–Weinberg genetic equilibrium (Cavalli-Sforza and Bodmer, 1971). Structure is the sum total of factors that disrupt genetic equilibrium, i.e., deviations from infinite population size (small or large-sized, population subdivision), departure from panmixis (assortative mating and inbreeding), and unequal contributions of specific phenotypes to the genetic makeup of the next generation (natural selection). Some population geneticists restrict the concept of genetic structure only to population subdivision, based on geography, language, religion, and ethnicity (Jorde, 1980). The relationships between the elements within populations, i.e. genes, genotypes, phenotypes, and groups of individuals have also been used to define the structure of human populations (Workman and Jorde, 1980). These approaches all assume that populations are the units of evolution. However, even the concept of population may be ignored and the pattern of distribution of genes may be considered for some analyses (Malecot, 1948).

Comparisons of populations or subpopulations are usually based on measures of genetic distance. Matrices of genetic distances among populations are often displayed through topological methods, such as principal component analyses (PCA) and phylogenetic trees. Further extension of PCA methodology has been used to generate synthetic gene maps, a method of studying the geographical distribution of genes (Cavalli-Sforza et al., 1994). However, Sokal et al. (1999) have demonstrated that this approach suffers from: (1) the inability to statistically test alternative hypotheses, and (2) possible misrepresentations associated with the extrapolation of missing gene frequencies used to create surfaces for the synthetic maps. Sokal and his colleagues instead have effectively utilized spatial autocorrelation techniques (SAC) to test hypotheses concerning gene distributions.

With the development of molecular genetic markers and their application to the reconstruction of human phylogeny, genetic structure studies of human populations have been reduced to comparisons of genetic distances, usually followed by the perfunctory dendrograms, constructed with the PHYLIP computer package. Some comparisons have been made within large human aggregates, such as documentation of caste gene flow, using mtDNA and Y-chromosome markers, among the castes of Andhra Pradesh, India (Bamshad *et al.*, 1998). However, the majority of studies in molecular genetics compare small, so-called random samples of populations with an eye to reconstructing the phylogeny or history of the groups. Few human populations practicing pastoral subsistence patterns, have been characterized genetically. Even less is known about the distribution of DNA markers in herding societies of Africa, the Americas and Asia. Several studies have been conducted among Basque populations (Spain and France), however, most of these groups were rural, but no longer follow a herding subsistence (Calderon, 2002). The Sámi reindeer herders of Finland, the Tuareg (flocks of goats, cattle, and camels) of the Sahara, and the Sinai Bedouins have been intensively studied genetically (Bonne *et al.*, 1971; Eriksson, 1973; Lefevre-Witier, 1982).

This chapter focuses on the genetic structure of two pastoral groups (the Evenki, of central Siberia and the Kizhi of Gorno Altai, southern Siberia). Although some genetic comparisons have been made among indigenous populations of Siberia, the formal population genetic analyses have been limited to the computation of genetic distances and the generation of phylogenetic trees (Torroni *et al.*, 1993; Starikovskaya *et al.*, 1998). Utilizing R-matrix analyses, some phylogenetic reconstructions have been based on Y-chromosome markers (Karafet *et al.*, 1999).

In this chapter, the reindeer herding (Evenki) and cattle-herding (Kizhi) populations are: (1) characterized on the basis of all of the available blood group markers, mitochondrial DNA (mtDNA), Y-chromosome markers, variable number tandem repeats (VNTRs), short tandem repeats (STRs), and an assortment of protein coding markers, such as alcohol dehydrogenase (ADH), aldehyde dehydrogenase (ALDH), and collagen 1A2 gene (COL1A2) – a DNA marker; (2) compared genetically to the surrounding groups, i.e., *interpopulation* analyses; (3) examined for the distribution of genes and the genetic structure within the herding aggregate or village, i.e., *intrapopulation* analyses.

These combined analyses should provide an adequate sampling of the genomes of Siberian populations and eliminate some of the biases observed in the reconstruction of human phylogeny. All too often, different regions of DNA or gene products yield vastly different population affinities or phylogenetic reconstruction. For example, the Y-chromosome haplotype analyses of the Siberian and Native American samples are suggestive of European ancestry (Santos et al., 1999) and have been trumpeted by archaeologists as proof of an Iberian connection with the Americas.

## Populations

### Evenki of Central Siberia

The Evenki are reindeer herders, widely distributed geographically throughout the taiga (birch and coniferous forests) of central Siberia, from the Ob-Irtysh watershed to the Sea of Okhotsk in the east. Vasilevich and Smolyak (1964) estimated that the Evenki occupied one-quarter of the entire territory of Siberia, 2.5 to 3 million square kilometers. Numerous river systems transect the Evenki territory. The swampy, sphagnum bogs and lichen-covered soils prevented any form of agriculture and necessitated hunting and herding subsistence economies of the Evenki. At the time of European contact (seventeenth century), the Evenki were continuously distributed from the Yenisey River to the eastern coast of Siberia, a distance of several thousand miles. However, this continuous distribution was interrupted by the rapid demic expansion of the Yakut from the south in the sixteenth century. According to the latest Russian census, a total of 29 900 Evenki continue to subsist in the taiga of Siberia (Figure 2.1).

The Evenki are speakers of the Tungusic language, which is a branch of the Altaic language family. The Evenki share this language family with Turkic, Mongolic, and Manchu-speakers (see a summary of the linguistic affiliations of Siberian indigenous populations in Table 2.1). The Manchu languages (spoken by Nanais, Orochi, Oroks, and Ul'chi) are linguistically most proximal to Tungusic and are classified as a subdivision of the Tungusic-Manchu branch of the Altaic language.

Data from Siberian archaeological excavations suggest that the Evenki were initially reindeer hunters, residing in the region of Lake Baykal, but that they expanded easterly and eventually adopted a herding economy (Okladnikov, 1964). Levin (1963) recounted an Evenki myth that they hybridized with a group from the east and acquired reindeer herding from them. Traditionally, the Evenki were organized into patrilocal and pat-

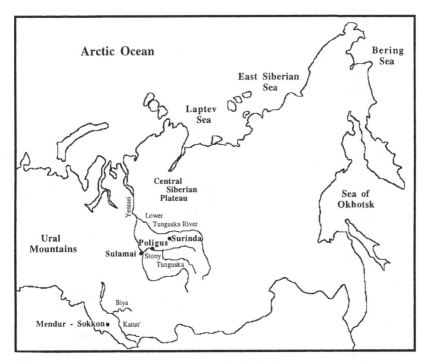

Figure 2.1. A map of Central Siberia, indicating the locations of the Evenki villages (Poligus and Surinda), the Ket village (Sulamai), and the Kizhi-Altai community (Mendur-Sokhon).

rilineal tribes, headed by princely families. Aside from the trading relationship with Russian merchants and the imposition of a fur tax, the Evenki were relatively isolated until the twentieth century. Their nomadic existence and mobility allowed them to escape any intrusions by Russian colonists into their traditional territories. This social isolation and organization was disrupted in the 1930s, when the Soviets tried to collectivize these reindeer herders. Many of the wealthy herd-owners (termed *kulaks* by the Soviets), and the traditional medicine men, (*Shamans*), were purged by the Soviet authorities. After massive executions, the reindeer herds were collectivized and the reindeer were tended by work groups organized by the collectives, and termed brigades (consisting of a melange of families and young, mostly unrelated males). These brigades replaced the extended families that traditionally herded the reindeer. Each cooperative group herds approximately 1500 head of reindeer, the size of the herd managed by an extended family or lineage during pre-Soviet times. After the breakup of the Soviet Union and the paucity of provisions from the outside, family groups have reclaimed most of the reindeer herds.

Table 2.1. *Linguistic affiliations of Siberian indigenous peoples*

| Language family | Ethnic group | Population size |
|---|---|---|
| *Altaic* | | |
| Turkic | Altais | 57 000 |
| | Tuvinians | 139 000 |
| | Tofalars | 720 |
| | Yakuts | 296 000 |
| | Dolgans | 6600 |
| Mongolic | Buryats | 314 000 |
| Tungusic-Manchu | | |
| Tungusic | Evens | 17 000 |
| | Evenks | 29 900 |
| | Udegeys | 1900 |
| Manchu | Orochi | 880 |
| | Ul'chi | 3200 |
| *Uralic* | | |
| Samoyedic | Entsi | 190 |
| | Nentsi | 34 200 |
| | Nganasans | 1300 |
| | Sel'kups | 3600 |
| Ugric | Khants | 22 300 |
| | Mansi | 8300 |
| *Paleoasiatic* | | |
| Chukotkan-Kamchatkan | | |
| | Chukchi | 15 100 |
| | Itel'mens | 2400 |
| | Koryaks | 8900 |
| Eskimo-Aleut | | |
| | Eskimos | 1700 |
| | Aleuts | 650 |
| *Linguistic Isolates* | | |
| | Kets | 1080 |
| | Nivkhs | 4600 |
| | Yukaghirs | 1100 |

### *Kizhi of Gorno Altai*

The Gorno Altai is geographically situated in Eurasia, lodged across China, Mongolia, Kazakhstan and Russian Siberia. The Gorno Altai region is inhabited by approximately 71 000 indigenous Turkic speakers and has experienced continuous human habitation for more than 50 000 years. This early habitation date is based on the discovery of Neandertal remains from the Okladnikov Cave of the Altai Mountains. During the past few thousand years, the Altai has been the crossroads between Europe and Asia, with major migrations and invasions emanating from both the East and the West. Mongols and Chinese dominated the region politically and remnants of gene flow and linguistic exchange can be detected in contemporary populations of the Altai. Okladnikov (1950) has documented the appearance of skeletal remains bearing European-like features from approximately 3500 BP. It has been hypothesized that Turkic-speaking peoples originated in the Altai and rapidly spread into adjoining regions (Bowles, 1977). However, Miller (1991) has argued that the Altaic language family developed in the west Siberian steppes more than 7000 years ago and spread to the Altai. In the twelfth century the Mongol hordes invaded the Altai and controlled the region politically until the eighteenth century, when the Oirat khanate began to crumble. By invading the southern Altai, the Chinese attempted to fill the political vacuum, caused by the disintegration of the society of the descendants of the western Mongols. The Russians forced the Chinese out of the Altai and assumed political control of the region in the latter part of the eighteenth century. In the 1930s, the Altaians were organized into farming and herding collectives by the Soviets (Forsyth, 1992).

Based on their language, culture and geographical distribution, the tribes of the Altai can be subdivided into two groups: (1) the Northern Altais, consisting of the Chelkanians, Tubalars, and Kumandinians; and (2) the Southern Altais, consisting of Altai proper, Kizhi, Telengits, Telesy, and Teleuts (Potapov, 1964). The Kizhi of the Gorno Altai Autonomous Republic are pastoral, semi-nomadic people who herd cattle on the grassy slopes of the surrounding mountains. The Kizhi are speakers of a Turkic language that is distantly related to Tungusic language, spoken by the Evenki.

### Genetics

In order to measure evolutionary change or phylogenetic affinities among groups, the gene pool of the populations must be sampled and characterized by typing DNA or protein markers. Gene frequencies or mutations can be calculated, identified, and genetic distances ascertained between populations. Although DNA can be extracted from buccal swabs, tissue, or hair, 7–10 ml of blood yields maximum information about the genetic makeup of a population. Blood samples yield: (1) significantly larger quantities of DNA than buccal swabs; (2) erythrocytes (red blood cells) can be tested for blood group antigens ABO, RH, MNS, Duffy, Diego, Kidd, Lewis, P, Lutheran, and XG*A; (3) red blood cell hemolysates can be phenotyped for proteins, such as hemoglobin, glucose-6-phosphate dehydrogenase (G-6-PD), phosphoglucomutase (PGM), acid phosphatase (ACP); (4) serum samples through immunotyping and electrophoresis, also yield considerable genetic information, through loci such as: immunoglobulins (GMs, KMs), albumin (AL), group-specific component (GC), haptoglobins (HP), transferrins (TF), and serum pseudocholinesterase (CHE).

## Population samples

Research was conducted during the summers of 1991, 1992, and 1993 among the indigenous populations of Siberia. In 1991, 75 blood samples were collected from volunteers of the village of Surinda, located along the Stony Tunguska River of Central Siberia (see Figure 2.1). This village has approximately 600 residents, including the 12 reindeer-herding brigades, located on territories surrounding the settlement. In 1992, the research team conducted fieldwork in Poligus, a village of approximately the same size and geographically adjacent to Surinda. A total of 44 samples were collected from Poligus and eight additional individual samples were obtained from Surinda. During the 1992 field season, a Ket village, Sulamai, located near the juncture of the Yenisey and Stony Tunguska rivers, was also sampled. In 1993, a joint Russian (Russian Academy of Sciences)–United States (University of Kansas) team studied a village of approximately 1000 residents, Mendur-Sokhon of Gorno Altai. A total of 95 persons from Mendur-Sokhon participated in this research program. The blood specimens collected in 1991–3, form the nucleus of the genetic analyses that are reported in this chapter.

Since few laboratories have the facilities and expertise to analyze all of the possible genetic markers, we collaborated with several research groups in order to broaden the sampling of the Siberian genetic variation. Thus, the Y-chromosome markers were tested by Chris Tyler-Smith's group, including Fabricio Santos, at Oxford University, and by John Mitchell of LaTrobe University in Australia. Similarly, the mitochondrial DNA was initially haplotyped by Tad Schurr of Douglas Wallace's laboratory at Emory University. John Mitchell's laboratory also analyzed some of the STRs, DAT1 and COL1A2 loci. The apolipoprotein variation was characterized through the collaboration with M. Ilyas Kamboh at the University of Pittsburgh. The VNTRs, using Southern blots and randomly amplified

Table 2.2. *Blood group frequencies of Evenki and Kets*

| System | Populations | | |
| --- | --- | --- | --- |
|  | Surinda | Poligus | Sulamai |
| *ABO*A* | 0.170 | 0.205 | 0.270 |
| *ABO*B* | 0.099 | 0.068 | 0.156 |
| *ABO*O* | 0.731 | 0.727 | 0.574 |
| *RH*R1* | 0.588 | 0.645 | 0.518 |
| *RH*R2* | 0.392 | 0.263 | 0.268 |
| *RH*r* | 0.016 | 0.092 | 0.214 |
| *RH*R1w* | 0.004 | 0.000 | 0.000 |
| *MNS*MS* | 0.087 | 0.079 | 0.093 |
| *MNS*Ms* | 0.547 | 0.526 | 0.518 |
| *MNS*NS* | 0.008 | 0.000 | 0.055 |
| *MNS*Ns* | 0.358 | 0.395 | 0.334 |
| *FY*A* | 0.843 | 0.763 | 0.667 |
| *FY*B* | 0.157 | 0.237 | 0.333 |

From M.H. Crawford *et al.*, (unpublished).

polymorphic DNA (RAPDs), were investigated at the Laboratory of Biological Anthropology, University of Kansas. Alcohol and aldehyde dehydrogenases were typed by Holly Thomasson and T.K. Li, Department of Medicine, Indiana University School of Medicine, Indianapolis. The blood group analyses and some of the protein markers for the Evenki were phenotyped at the Minneapolis War Memorial Blood Bank, under the direction of Herbert Polesky. The remainder of the protein markers and the immunoglobulins were tested by the Analytical Genetic Testing Laboratory of Denver, under the direction of Moses Schanfield. Currently, James Knowles, at Columbia University, is characterizing Siberian populations with more than 300 STR loci, distributed over the genome at 10 centiMorgan distances.

### Standard markers

#### Blood groups

Although the Altai population (Mendur-Sokhon) was not characterized for the blood group antigens, two Evenki communities (Surinda and Poligus) and one Ket village (Sulamai) were tested for the ABO, RH, MNS and Duffy blood groups (see Table 2.2). A complete analysis of blood group gene frequencies and phenotypic frequencies, plus their compari-

Table 2.3. *Red blood cell protein variation among the herders of Siberia*

| System | Populations | | | |
|---|---|---|---|---|
| | Surinda (Evenki) | Poligus (Evenki) | Sulamai (Kets) | Mendur-Sokhon (Kizhi-Altai) |
| *ACP1* | | | | |
| *A | 0.322 | 0.343 | 0.141 | 0.414 |
| *B | 0.668 | 0.643 | 0.652 | 0.581 |
| *C | 0.009 | 0.014 | 0.000 | 0.005 |
| | | | | |
| *ESD* | | | | |
| *1 | 0.865 | 0.878 | 0.724 | 0.742 |
| *2 | 0.135 | 0.122 | 0.276 | 0.253 |
| *5 | 0.000 | 0.000 | 0.000 | 0.005 |
| | | | | |
| *PGM1i* | | | | |
| *1A | 0.767 | 0.526 | 0.679 | 0.532 |
| *1B | 0.036 | 0.171 | 0.179 | 0.113 |
| *2A | 0.193 | 0.303 | 0.089 | 0.269 |
| *2B | 0.004 | 0.000 | 0.053 | 0.086 |

From M.H. Crawford *et al.*, (unpublished).

sons with other Siberian populations is included in Crawford *et al.*, (Genetic markers in Siberian indigenous populations, unpublished, 2000). The frequency of *ABO*O* in both of the Evenki villages is somewhat higher than the frequencies observed among other Siberian reindeer herders and hunters, such as the Nentsi, reindeer Chukchi, and the Nganasans of Siberia. The two adjoining Evenki communities resemble each other genetically for most of the blood group loci. However, significant difference was observed for the *RH*R2* (*Cde*) haplotype, 0.39 versus 0.26. These small differences in the allelic frequencies between the two Evenki villages are most likely due to the small sample size for Poligus and familial overrepresentation in the samples, because of the small population sizes of the herding communities.

### Erythrocytic protein markers

Table 2.3 summarizes the allelic frequencies of three erythrocytic protein loci: ACP1, esterase D (ESD) and phosphoglucomutase-1 (*PGM1*) in Evenki, Altai and Ket populations. The ACP and ESD loci were phenotyped using standard electrophoretic procedures, while the PGM1

Table 2.4. *Serum protein variation in herding and hunting populations of Siberia*

|  | Populations | | | |
|---|---|---|---|---|
| System | Surinda | Poligus | Sulamai | Mendur-Sokhon |
| *GC* | | | | |
| *1 | 0.888 | 0.972 | 0.704 | 0.837 |
| *2 | 0.112 | 0.028 | 0.296 | 0.163 |
| *F13A* | | | | |
| *1 | 0.909 | 0.918 | 0.896 | 0.817 |
| *2 | 0.082 | 0.091 | 0.063 | 0.183 |
| *V | 0.000 | 0.000 | 0.042 | 0.000 |
| *F13B* | | | | |
| *1 | 0.448 | 0.321 | 0.604 | 0.484 |
| *2 | 0.006 | 0.012 | 0.000 | 0.022 |
| *3 | 0.547 | 0.667 | 0.396 | 0.467 |
| *Altai1 | 0.000 | 0.000 | 0.000 | 0.016 |
| *Altai2 | 0.000 | 0.000 | 0.000 | 0.011 |
| *PLG* | | | | |
| *A | 0.994 | 0.971 | 1.000 | 0.940 |
| *B | 0.006 | 0.029 | 0.000 | 0.060 |

From M.H. Crawford *et al.*, (unpublished).

locus was characterized by isoelectric focusing (IEF). As expected, Sulamai and Mendur-Sokhon differ genetically on the basis of their gene frequencies from the two Evenki communities. However, there is considerable difference between Surinda and Poligus on the basis of PGM1 locus with the *1A* allele occurring at a frequency of 77% in Surinda versus 53% in Poligus. Similarly, there are differences in *PGM1\*2A* (19.3% versus 30.3%) and *1B* (4% versus 17%) between the two adjacent Evenki communities. In the other two loci, ACP1 and ESD, Surinda and Poligus are very similar in allelic frequencies (Surinda = 32% and Poligus = 34%) but both differ significantly from Sulamai and Mendur-Sokhon, Sulamai = 14% and Mendur-Sokhon = 41%.

### Serum proteins

Table 2.4 summarizes the allelic frequencies of four serum protein loci (GC, F13A, F13B, and PLG) in Evenki, Altai and Ket populations. A more complete analysis of a larger set of loci for these populations is contained in

Crawford *et al.* (unpublished). With the exception of plasminogen (PLG), there is considerable variation exhibited by the other serum protein loci. A number of private or rare alleles are displayed in the F13A and F13B in the Altai and in Sulamai. One rare allele, *F13A\*V* appears only in Sulamai at a frequency of 4%. Two new, possibly familial rare alleles, *Altai1* and *Altai2* (0.016 and 0.011) respectively, were observed only in Mendur-Sokhon. It is surprising to detect so many variants in a system such as fibrinogen, which plays such a critical function in blood clotting.

### DNA markers

#### Mitochondrial DNA

The research on mitochondrial DNA (mtDNA) variation resulted from collaboration with Douglas Wallace, Antonio Torroni, and Tad Schurr of Emory University. The initial results (Torroni *et al.*, 1993) were further extended by Schurr *et al.*, 1999, through the use of high-resolution restriction fragment length polymorphisms (RFLP) and sequencing analyses.

mtDNA differs from nuclear DNA in a number of ways: (1) it is found in the mitochondria of the cytoplasm; (2) it is a duplex, circular DNA molecule of 16 400 base pairs (bp) in length; (3) it consists primarily of coding sequences; (4) it is inherited maternally and thus reflects maternal lineage and female migration; (5) apparently, it does not experience recombination and as a result, the only genetic change is due to an accumulation of mutations at a rate of 2–4% per million years. However, recent analyses of linkage disequilibrium are suggestive of possible recombination of hominid mtDNA (Awadalla *et al.*, 1999). If the interpretations of the data by Adwadalla and colleagues are accurate, mtDNA-based clocks would have to be re-evaluated and corrected for possible recombination. Torroni *et al.* (1993) had argued that the date of divergence of Siberians from Native Americans on the basis of mtDNA was 17 000–34 000 BP. However, these dates are in question given the latest evidence that the observed variation may not be entirely due to mutation buildup and that some recombination is necessary to explain the observed patterns of linkage disequilibrium (Awadalla *et al.*, 1999).

A total of 15 mtDNA haplotypes (defined by the presence or absence of restriction sites and/or deletions) were observed within the Evenki sample of 51 persons (Torroni *et al.*, 1993). The most common haplotypes all belonged to the *C* haplogroup *S26* (21.6%), *S29* (19.6%), *S27* (11.8%) and *S32* (9.8%). Of the populations listed in Table 2.5, only the Northern Altai

Table 2.5. *Mitochondrial DNA haplogroups in Siberian populations using high-resolution restriction (RFLP) analysis and control region (CR) sequencing*

| Population | Sample (n) | % of mtDNA Haplogroups | | | | |
|---|---|---|---|---|---|---|
| | | A | B | C | D | Other |
| Evenks | 51 | 3.9 | 0.0 | 84.3 | 9.8 | 2.0[a] |
| Udegeys | 45 | 0.0 | 0.0 | 19.6 | 0.0 | 80.2[b] |
| Nivhks | 57 | 0.0 | 0.0 | 0.0 | 28.1 | 71.0[c] |
| N. Altai | 28 | 3.6 | 3.6 | 35.7 | 14.3 | 42.8[d] |
| Kets | 23 | 0.0 | 0.0 | 21.7 | 0.0 | 78.3[d] |

[a] The "Other" category consists of Γ-haplogroups.
[b] The 80.4% of previously unidentified haplogroups consisted of Y (8.9%), I (28.9%), and III (44.4%) haplogroups.
[c] The 71% of "Other" haplogroups were identified as Y (65%) and I (1.8%).
[d] These populations were described by Sukernik *et al.*, (1996) *Russian Journal of Genetics*. The northern Altai populations are mainly Chelkanians and Tubalars.
From Schurr *et al.* (1999).

Group exhibits the *B* haplogroup, present in several geographical regions of the Americas, Mongolia, and coastal Asia. Interestingly, the 9 bp deletion, characteristic of the *B* haplogroup, is absent in Siberian populations most proximal to the Americas, i.e. the indigenous groups of Chukotka, Kamchatka, and other northeastern regions of Siberia. Only the Mongols share this haplogroup with the Altai and the Americas.

### Y-chromosome markers

Most of the Y-chromosome is paternally inherited and haploid thus escaping recombination. The accumulated mutations on this portion of the Y, therefore, are a record of its evolutionary history, and can be used to investigate male-specific aspects of human population structure and variation. Further, Y-chromosome polymorphisms exhibit much greater geographical variation than other genetic (autosomal and mitochondrial DNA) polymorphisms. This is principally due to the small effective size of the Y-chromosome and further exaggeration due to cultural practices such as polygyny. The Y chromosome is of great value because it contains different types of polymorphisms; some of which result from either unique or rare substitution events, while others, such as microsatellites are subject to more mutational events.

The genetic characterization of Y-chromosome markers for the Kets, the Evenki, and the Altai populations is principally a product of recent and ongoing research collaboration with C. Tyler-Smith, F. Santos, and J. Mitchell. To date, seven polymorphic systems were scored, of which six were the rare event markers and the other was the complex and more variable alphoid heteroduplex ($\alpha$h) system (Santos et al., 1999). These markers generated a variety of Y haplotypes in the three Siberian groups. The most common haplotype in the Evenki (18/23) is not present in the other two groups, but occurs among Mongolians, and is clearly an Asian chromosome. Of the other four haplotypes seen in Evenki, two are distinctly Asian haplotypes. The Altai exhibited six unique Y haplotypes, with one characterized by $\alpha$h type II and the SRY-1532 A allele by far the most common (12/23). This haplotype is also common in Kets (2/10) and present in the Evenki (1/23) and has been reported in Europeans, Indians, and to a lesser extent in Mongolians. The Altai (4/23) and Ket (7/10) also share a relatively high frequency of a haplotype characterized by type $\alpha$h I and 92R7 T allele. Most interestingly, this chromosome, so far, has only been detected in native Americans and also in a Mongolian. Two haplotypes shared by the Evenki and the Altai ($\alpha$h type XVIII and $\alpha$h type XX respectively) are also present in Mongolians. One Ket has a haplotype that provides indirect evidence of admixture with a European source, most likely from Russia. In a separate study (Shinka et al., 1999) of the distribution of a Y haplotype found in Japan and Korea (known as haplotype x') no examples were found in any Siberian population, including the three groups of the present study.

The overall picture of Y-chromosome variation among the three groups can be summarized in a measure known as genetic diversity. Genetic diversity levels in these pastoralists ranged from a low of 0.374 in the Evenki to a maximum of 0.661 in the Altai, with the Kets having an intermediate value (0.460). These findings are consistent with our knowledge of their history, demography and ecology. The Evenki are the most isolated of the groups geographically and live in small herding units and in small villages dispersed across the Siberian taiga. They show minimal evidence of admixture with Russians and their closest affinities are with Mongolians.

The increased diversity in the Altai, however, appears to reflect this population's position at the crossroads of Asia and Europe, and accordingly they experienced gene flow from Mongolian and Turkic groups. Interestingly, though the Tat polymorphism, which defines Y-haplogroup 16 (Zerjal et al., 1997) is present in some members of the Altaic language family (Mongolians, Buryats and Yakuts), it is absent from the three groups. Present work on Y-chromosome markers is focusing on examining diversity within each of the haplotypes or lineages through the analysis of associated microsatellite polymorphisms (Mitchell *et al.*, unpublished).

### Non-coding DNA

#### VNTRs

VNTRs (variable number of tandem repeats) or minisatellites are stretches of non-coding DNA in which nucleotide sequences 20–50 bp in length are repeated tandemly from 10 to 100 times. They appear throughout the human genome and exhibit high mutation rates. Jeffreys *et al.* (1985) first described these hypervariable minisatellites and adapted the multilocus VNTRs for use as "fingerprints" because the copy number varies to such a degree that each individual has a unique configuration.

Siberian populations were compared with each other and with Native Americans, European, and African American groups on the basis of single locus VNTR DNA fragment frequency distributions. VNTRs were used in this study because: (1) they are believed to be neutral evolutionarily, i.e. better suited for studying genetic drift and population structure; (2) the observed high mutation rates provide a focus on recent human evolution (Harding 1992).

DNA was extracted, restricted with PstI enzyme, and the fragments were separated electrophoretically on agarose gels. The separated DNA fragments were transferred onto a nylon membrane and hybridized with five different biotinylated probes (D7S104, D11S129, D18S17.1, D20S15, and D21S112 loci) available from Collaborative Research Incorporated (CRI). The VNTR fragments were digitized using molecular sizing ladders and grouped into ±2% error bins (for methodological detail see McComb *et al.*, 1996). Distributions of the binned DNA fragments were compared using the Kolmogorov–Smirnov test protected by the Bonferroni method for chance significance arising from multiple comparisons. The distribution of DNA fragments for the Kets do not differ significantly from those of the Altai population. However, the Evenki differ significantly from both the Altai and the Kets for some of the loci (see Table 2.6). The Siberian populations differ significantly in their VNTR fragment distributions when compared with African-Americans, Europeans, and, to a lesser degree, Native Americans.

Multivariate analyses, based on these five VNTRs, are described in the

Table 2.6. *Kolmogorov–Smirnov test results for the distribution of variable number tandem repeat ( VNTR) fragments within four Siberian populations*

|  |  | Kets (Sulamai) | Evenki (Surinda) | Evenki (Poligus) |
|---|---|---|---|---|
| Altai | D7S104 | 0.21 | 0.54 | 0.31 |
| (M-S)[a] | D11S129 | 0.45 | 0.51* | 0.39 |
|  | D18S17 | 0.25 | 0.27 | 0.48* |
|  | D20S15 | 0.12 | 0.30* | 0.27 |
|  | D21S112 | 0.25 | 0.42 | 0.34 |
| Kets | D7S104 |  | 0.57 | 0.14 |
| (Sulamai) | D11S129 |  | 0.71* | 0.71* |
|  | D18S17 |  | 0.16 | 0.28 |
|  | D20S15 |  | 0.32 | 0.35 |
|  | D21S112 |  | 0.21 | 0.16 |
| Evenki | D7S104 |  |  | 0.67 |
| (Surinda) | D11S129 |  |  | 0.24 |
|  | D18S17 |  |  | 0.32 |
|  | D20S15 |  |  | 0.28 |
|  | D21S112 |  |  | 0.19 |

* Significant with Bonferroni Protection ($p < 0.05$).
[a] Mendur-Sokhon.
From McComb *et al.*, (1966).

Analytical Procedures section of this chapter. R-matrix analyses, phylogenetic dendrograms, and discriminant analyses demonstrate the usefulness of VNTRs in the reconstruction of human evolution.

### *DAT1*

DAT1 is a VNTR polymorphism in the 3' untranslated region of human dopamine transporter gene. This VNTR consists of a 40-bp core sequence tandemly repeated from 3 to 13 times with the number of repeats being equivalent to alleles. A recent survey of 10 human populations from Africa, Asia, Europe, the Americas, and Australia, revealed the presence of five DAT1 alleles (Mitchell *et al.*, 2000). Four Siberian populations (Kizhi from the Altai, the Evenki and Kets from the Stony Tunguska river, and the Sel'kup from north-central Siberia) were tested for DAT1 variation. However, only the population from the Altai exhibited all five of the

DAT1 alleles, *DAT\*7* (360 bp), *9* (440 bp), *10* (480 bp), *11* (520 bp), and *13* (600 bp). In some population samples, e.g. Native Americans of Colombia, *DAT1* is monomorphic with only *DAT1\*10* being present. The usually very rare *DAT1\*7* allele occurs in all four of the Siberian populations at approximately 5%. Another rare allele *DAT1\*13* was observed only in Mendur-Sokhon, and Mongolia. It is not surprising that these two populations share the *DAT1\*13* allele, given the invasion of the Altai by the Mongols and continued cultural contact.

This particular VNTR locus provides useful historical and evolutionary information, as it appears to have few alleles and mutate slowly. Rare mutations, such as *DAT1\*13* or *DAT1\*7*, reveal population migrations and affinities. For example, the *DAT1\*7* allele is restricted to Siberia and Australian Aborigines. Yet this allele is absent among the Europeans, and, therefore, could not have been introduced through European contact into Australia. Mitchell *et al.* (2000) suggested that this allele was introduced into Australia by admixture with Indonesian fishermen, or Chinese traders or Japanese pearl collectors. Thus, unexpected pathways of human contact can be traced using these infrequent genetic variants.

### Short tandem repeats

Microsatellite or short tandem repeat (STRs) loci consist of tandemly repeated sequences of non-coding DNA containing a core sequence of between 1 and 5 bp and with the total repeat unit usually less than 350 bp. There are some limitations to the use of microsatellites with respect to population and evolutionary studies. Because STRs display relatively high mutation rates and the mechanism of mutation is thought to be single mutation by replication slippage, identical alleles can occur in different populations which have no recent shared ancestry. Consequently, the more recent the divergence between populations, the more likely it is that shared STR alleles are related by descent. Microsatellite markers are particularly well suited to investigations of recently diverged groups (i.e., less than 50 000 years ago) and the Siberian pastoralists fall well within this range.

We scored four STRs/microsatellites on the Evenki, Ket and Kizhi-Altai (see Table 2.7), as well as Native Americans from the United States, Europeans, Javanese (from Surabaya) and Asians (principally Vietnamese and southern Chinese Han). The microsatellites TPOX, CSF1PO, THO1 and vWA have been scored on many human groups as a result of their applications in forensic science. All four of these loci are tetranucleotide polymorphisms and all have simple repeat structures (a single type of 4-bp

Table 2.7. *Short tandem repeat (STR) variation in Siberian populations*

| Loci (Alleles) | Populations | | |
|---|---|---|---|
| | Evenki | Kets | Altai |
| *THO1* | | | |
| 6 | 0.018 | 0.063 | 0.087 |
| 7 | 0.282 | 0.375 | 0.320 |
| 8 | 0.436 | 0.125 | 0.053 |
| 9 | 0.127 | 0.281 | 0.373 |
| 9.3 | 0.136 | 0.156 | 0.160 |
| 10 | 0.000 | 0.000 | 0.007 |
| *vWA* | | | |
| 14 | 0.136 | 0.115 | 0.059 |
| 14.2 | 0.000 | 0.039 | 0.000 |
| 15 | 0.045 | 0.077 | 0.088 |
| 15.2 | 0.045 | 0.039 | 0.088 |
| 16 | 0.159 | 0.039 | 0.235 |
| 16.2 | 0.000 | 0.000 | 0.029 |
| 17 | 0.250 | 0.308 | 0.412 |
| 18 | 0.318 | 0.154 | 0.088 |
| 19 | 0.045 | 0.231 | 0.000 |
| *CSF1PO* | | | |
| 9 | 0.017 | 0.045 | 0.016 |
| 10 | 0.219 | 0.454 | 0.286 |
| 11 | 0.202 | 0.136 | 0.349 |
| 12 | 0.526 | 0.364 | 0.286 |
| 13 | 0.026 | 0.000 | 0.055 |
| 14 | 0.009 | 0.000 | 0.008 |
| *TPOX* | | | |
| 8 | 0.368 | 0.545 | 0.484 |
| 9 | 0.044 | 0.045 | 0.159 |
| 10 | 0.017 | 0.000 | 0.008 |
| 11 | 0.386 | 0.364 | 0.230 |
| 12 | 0.088 | 0.045 | 0.119 |
| 13 | 0.096 | 0.000 | 0.000 |

From Mitchel *et al.*, (2000).

repeat), except vWA that has a compound structure. The number of alleles at these loci range from 8 (TPOX) to 15 (vWA).

At the THO1 locus the Kizhi-Altai and Ket show the greatest similarity, with the Evenki displaying some very distinct allelic frequencies. In particular, the frequency of allele 8 in the Evenki (43%) distinguishes them from all other world populations (where the frequency is < 10%). They are also distinctive in their very low incidence of allele 6. The Altai and Kets show most similarity to Asians especially in the frequencies of alleles 6, 7, and 9.

At TPOX the Evenki and Ket show the greatest similarity, with the Altai having a high frequency of allele 10 and low incidence of 11. The Evenki are distinguished by a high frequency of allele 13 which is absent in the other two groups.

At the CSF1PO locus the Evenki have a remarkably high frequency of allele 12 (53%), whereas it is only 29% in the Kizhi-Altai, and the Kets have a very high incidence of allele 10 (45%).

The vWA locus exhibits the irregular/intermediate repeat alleles, 14.2, 15.2, and 16.2 are all seen in the Siberian groups and some in relatively high frequency. The other alleles differ widely across the three groups; for example, allele 18 ranges from 8% in the Kizhi-Altai to 31% in the Evenki, and allele 19 from 0 in the Kizhi-Altai to 23% in the Ket. Some of these fluctuations in the frequencies of the alleles were due to relatively small sample sizes, as with the Kets.

### Coding DNA

#### ADH and ALDH

Variation in the metabolism of alcohol may play a significant role in predisposing individuals to alcoholism, a condition of high prevalence in Siberian indigenous populations (Crawford, 1998). Alcohol is metabolized by hepatic alcohol dehydrogenase (ADH) to acetaldehyde and by mitochondrial aldehyde dehydrogenase (ALDH2) to acetate. There is considerable variation at the ADH2 and ADH3 loci that produce isozymes with differing catalytic properties. The ADH2 and ADH3 loci encode for more active isozyme subunits that are found in high frequencies among Asian populations. Table 2.8 summarizes the frequencies of ADH and ALDH alleles in Siberian populations. *ADH2\*2* allele produces a high-activity enzyme, while *ALDH2\*2* allele results in a low enzyme activity and a greater propensity for the flushing response. *ADH2\*2* allele varies from less than 1% among the Kizhi-Altai to 23% among the Buryats. The *ALDH2\*2* allele, absent or at low frequencies in Native American and Siberian populations, occurs at a higher frequency among the Nivkhi. The section of this chapter on Analytical Procedures contains a discussion of the application of ADH and ALDH to the reconstruction of human evolutionary history (see Figures 2.7 and 2.8).

Table 2.8. *Frequencies of alcohol dehydrogenase (ADH) and aldehyde dehydrogenase-2 (ALDH2) alleles in Siberian populations*

| Populations | ADH2 locus | | | ADH3 locus | | ALDH2 |
| | ADH2*1 | ADH2*2 | ADH2*3 | ADH3*1 | ADH3*3 | ALDH2*2 |
| --- | --- | --- | --- | --- | --- | --- |
| Evenki | 90 | 7 | 3 | – | – | 0 |
| Nivkh | 91 | 9 | 0 | – | – | 9 |
| Buryats | 77 | 23 | – | 83 | 17 | 0 |
| Altai | 99 | < 1 | – | 19 | 81 | 0 |

From Thomasson *et al.*, (2001).

### Apolipoproteins

Apolipoproteins are the protein components of plasma lipoprotein particles that play a significant role in the secretion, processing and breakdown of lipoproteins. Much attention has been paid to the apolipoproteins because of their role in regulating plasma lipid levels and potential risks of cardiovascular disease. Seven polymorphic sites in three apolipoprotein genes (APOE, APOH, and APOA4) were studied among the Evenki of Surinda and Poligus (Kamboh et al., 1996). In addition, total cholesterol, triglycerides, high density- and low density-lipoproteins were measured from serum specimens. Unfortunately, no comparative data are available for either the Kets or the Kizhi-Altai.

APOE is the most polymorphic gene involved in lipoprotein metabolism. It serves as a ligand for two cell receptors, low-density lipoprotein particles (LDL) and LDL-related protein receptor (LRP). APOE mediates the uptake of apoE-containing lipoprotein particles by the cells. Three common alleles are usually found: *APOE*2*, *APOE*3*, and *APOE*4*.

The allelic frequency data reveal several interesting features among the Evenki: (1) the near absence of the *APOE*2* allele (a unique marker for Europeans); (2) the highest-ever recorded frequency of the *APOH*3* allele; (3) sporadic incidence of variant alleles at codons 130 and 347 in the *APOA4* gene.

The well-established association between the *APOE*4* allele and elevated plasma levels in European populations was not observed among the Evenki, despite their having comparable frequency of the *APOE*4* allele (Kamboh et al., 1996). Evenki women with the *APOE*4* allele exhibit significantly lower LDL cholesterol levels when compared with those who

Table 2.9. *Distribution of APO\*E, APO\*A and APO\*H alleles among the Evenki*

| Alleles | Frequencies |
|---|---|
| *APOE* | |
| *2 | 0.004 |
| *3 | 0.843 |
| *4 | 0.153 |
| | |
| *APOH* | |
| *1 | 0.013 |
| *2 | 0.788 |
| *3 | 0.199 |
| | |
| *APOA4* | |
| Codon 127 (Hinc II) | + = 0.707 |
| | − = 0.293 |
| Codon 130 (Hae III) | + = 0.993 |
| | − = 0.007 |
| Codon 347 (Hinf I) | A = 0.984 |
| | T = 0.016 |
| VNTR (exon 3) | D = 0.289 |
| | I = 0.711 |

From Kamboh *et al.*, (1996).

have the *APOE\*3* allele. Kamboh *et al.* (1996) hypothesize that the apparent absence of association of the *APOE\*4* allele to increased levels of lipids may be due to a gene–diet interaction that modulates the effect of the APOE polymorphism (Table 2.9).

## COL1A2

Rare mutations, particularly deletions, and insertions, can sometimes be utilized to reconstruct unique evolutionary events and the common ancestry of specific populations. One deletion of 38 bp in an intron within the human α2 (1) collagen gene (COL1A2) was found among the Evenki (0.34), Kets (0.33), and Altai (0.18) populations (Mitchell *et al.*, 1999). The high frequency of this deletion of a sequence (found only in *Homo sapiens*) and its absence in sub-Saharan Africa, suggests that the deletion event occurred either prior to or shortly after modern humans left Africa (see Table 2.10). All of the non-African populations tested to date exhibit the

Table 2.10. *Frequency of the COL1A2 deletion (603 bp) in human populations distributed throughout the world*

| Populations | Numbers sampled | Frequency of deletion |
|---|---|---|
| *Africa* | | |
| Benin, Bariba | 32 | 0.00 |
| Benin, Dendi | 50 | 0.00 |
| Biaka Pygmy | 45 | 0.00 |
| Mbuti Pygmy | 53 | 0.00 |
| San | 7 | 0.00 |
| | | |
| *Europe* | | |
| England | 35 | 0.34 |
| Italy, Calabria | 74 | 0.28 |
| Sardinia | 76 | 0.32 |
| Norway | 180 | 0.34 |
| Chuvash | 34 | 0.28 |
| | | |
| *Asia* | | |
| Sel'kup | 17 | 0.24 |
| Evenki | 16 | 0.34 |
| Ket | 9 | 0.33 |
| Altai | 53 | 0.18 |
| | | |
| *Americas* | | |
| US Native Americans | 19 | 0.29 |
| Colombian Native Americans | 27 | 0.22 |
| Cayapa, Ecuador | 38 | 0.34 |
| | | |
| *Oceania* | | |
| Australian Aboriginals | 17 | 0.18 |
| Solomon Islands | 31 | 0.00 |

From Mitchell *et al.*, (2000).

deletion allele at a relatively high frequency. In contrast, Africans without European admixture lack this COL1A2 deletion. The same pattern of geographical distribution as observed in the COL1A2 deletion was found in two other loci, the $GM*A, X G$ haplotype and a C→T mutation in the human $\beta$-pseudogene of hemoglobin. $GM*A, XG$ is observed in all human populations tested thus far, except for sub-Saharan Africans. Hominids coming out of Africa must have formed a small population that experienced several mutations that were later carried to all human populations who descended from them.

The incidence of the COL1A2 deletion in Siberian populations yields little new information about the affinities of the various indigenous groups.

Evenki and Kets have the highest frequency of the deletion, with the Sel'kups (a Samoyedic group) exhibiting intermediate frequencies. The Kizhi-Altai has the lowest incidence of the deletion in the Siberian populations tested to date.

What does this distribution of collagen genes and immunoglobulin haplotypes tell us about human evolution? Certainly, these data are consistent with the out-of-Africa hypothesis of humans sometime between 60 000 and 120 000 years ago. In addition, these data reveal that there was no back migration to Africa. Lastly, the presence of three mutations in all populations out of Africa suggest that the group that migrated out must have paused for some time before dispersal to all corners of the world.

### Analytical procedures

#### *Inter-population analyses*

Genetic similarities and differences among populations or subdivisions are measured by the frequencies of alleles or haplotypes in human populations. These frequencies can be summarized in the form of genetic distances between populations. However, large-scale numerical comparisons between many populations are difficult to visualize, thus, various graphic displays based on distance matrices have been developed (Crawford, 1998). These displays can roughly be subdivided into phylogenetic trees (dendrograms), and topological representations. One of the most commonly utilized topological approaches to the study of population affinities is principal components analysis of a variance-covariance R-matrix (Harpending and Jenkins, 1973). The variance-covariance matrix of allelic frequencies is deconstructed into eigenvectors that correspond to the proportion of variation observed in the matrix. Populations with similar frequencies are clustered in plots of the scaled eigenvectors (see Figure 2.2).

#### *VNTRs*

Figure 2.2 is a principal components plot using the frequencies of binned DNA fragments from five VNTR loci (D7S104, D11S129, D18S17, D20S15, and D21S112). The methodology used in this analysis has been

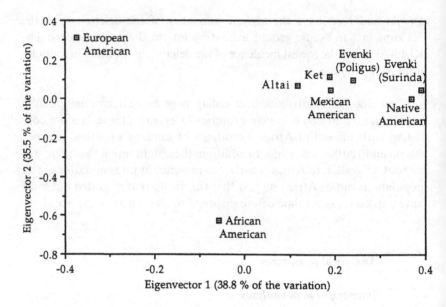

Figure 2.2. Principal components analysis of the Siberian and Native American R-matrix, based on five variable number tandem repeat (VNTR) loci (McComb *et al.*, 1995).

described in McComb *et al.*, 1995, 1996). The Kets, Evenks, and Altai populations of Siberia cluster with Native American and Mexican American samples. European American and African American samples are the most distinctive and do not cluster with any other groups. The Altai, Kets, and Mexican Americans have a slight tendency towards the Europeans, while the Native Americans and Surinda Evenks form a tight cluster that represents little or no admixture.

Figure 2.3 is a plot of the genetic distance from the centroid of distribution versus the mean per locus heterozygosity based on the five VNTR loci. A theoretical regression line has been fitted to the points using the method of Harpending and Ward (1982). Populations experiencing admixture or high levels of migration and gene flow exhibit high mean per locus heterozygosity and low levels of $r_{ii}$ distance from the centroid of distribution. Populations experiencing genetic isolation and the effects of stochastic processes are predicted to have high $r_{ii}$ and low heterozygosity. Poligus, an Evenki settlement, exhibits the highest $r_{ii}$ level, followed by Surinda. These two Evenki populations are the most geographically and reproductively isolated groups that have differentiated from other Siberian populations. The Altai community (Mendur-Sokhon) and the Ket village (Sulamai),

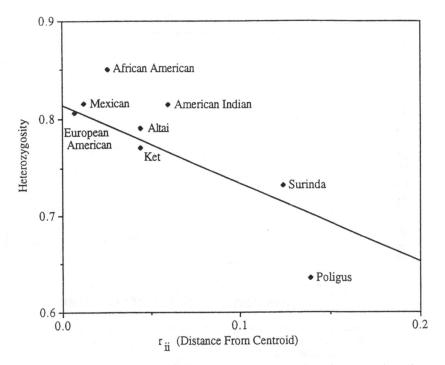

Figure 2.3. A plot of the relationship between mean per locus heterozygosity and the distance from the centroid of distribution ($r_{ii}$) (McComb *et al.*, 1995).

although both communities are highly isolated geographically, have both experienced considerable gene flow from Russians and Mongols.

The $G_{st}$ value for Siberia, based upon five VNTR loci, is 0.032 (significant at $p < 0.002$, when compared with a random distribution of 500 values). This high $G_{st}$ value is in part due to the small size of these populations and their genetic isolation (McComb *et al.*, 1996).

### STRs

Figure 2.4 is a plot of the first two components of the relationship (R) matrix based on STR frequencies for the four loci of the three Siberian populations (Evenki, Altai, and Kets) shown in Table 2.6 with four other populations (Java, Asia, Caucasian, and Native American). This plot accounts for 57% of the total variation. The first component separates the Evenki from all of the other groups. The second component separates the two Asian populations from the Europeans, Native Americans, and the

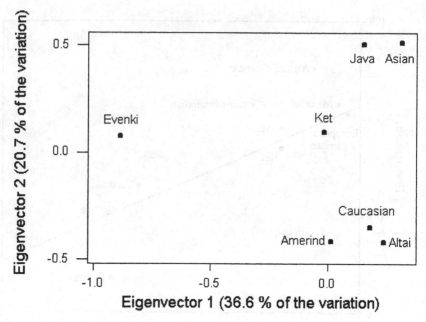

Figure 2.4. Principal components analysis, based on four short tandem repeat (STR) loci of Siberian populations compared with Native Americans, Europeans, and Asians.

Kizhi-Altai community. The Evenki and the Ket have an intermediate position along the second dimension. The Kets plot close to the center of the distribution of these populations, thus, revealing some admixture with Russian settlers.

Figure 2.5 is a dendrogram (plotted by the Fitch–Margoliash method with contemporary tips, version 3.573c) of the same seven populations on the basis of STR frequencies used in the R-matrix gene map shown in Figure 2.4. The Shriver *et al.*, 1995, method was used for the calculation of the genetic distances employed. The Evenki are the most distinctive of the seven populations and branch off the tree at the greatest genetic distance. The Kizhi-Altai branch bifurcates next from the combined Siberian, Native American and European branch. The Europeans cluster most closely with the Native Americans while the Javanese cluster with the Asians.

The plot of mean per locus heterozygosity versus $r_{ii}$ based on STRs is shown in Figure 2.6. Only the Kets and the Europeans are outside the 95% interval. The Kets have the lowest mean per locus heterozygosity but an average $r_{ii}$ level. The village of Sulamai is located near the confluence of Stony Tunguska and the Yenesey Rivers and is isolated geographically

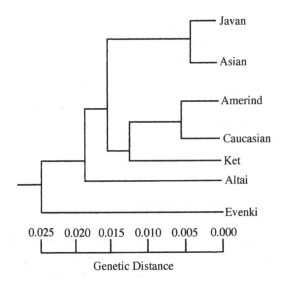

Figure 2.5. Phylogenetic dendrogram based on STRs for Siberian, Asian, Native American and European populations. Genetic distances were computed using Shriver *et al.*, (1995) method, and cluster analysis utilized the Fitch–Margoliash method.

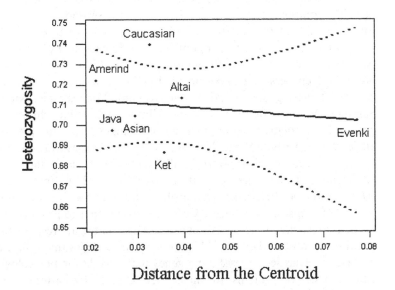

Figure 2.6. A plot of the relationship between mean per locus heterozygosity (H) and the distance from the centroid of distribution ($r_{ii}$) for four STR loci. The dotted line represents 95 percentiles in the distribution.

from the closest villages. Yet, Russian settlers now constitute about one half of the community. The Evenki have the highest $r_{ii}$ of any of the populations in this comparison. Yet, the Evenki heterozygosity level is higher than that of the Kets, Java, and the Asian samples. When compared with a similar plot based upon VNTRs (see Figure 2.3) the Evenki (sub-divided into two villages) continue to show high $r_{ii}$ levels. However, there is less differentiation of the populations using STRs than when VNTRs are utilized to compare populations.

### Protein coding regions of the genome

Figure 2.7 contains an R-matrix analysis plot based on allelic frequencies of ADH and ALDH loci and representing Siberian, Asian, and Oceanic populations. Unlike the non-coding VNTRs and STRs, ADH and ALDH play an integral part in the metabolism of an organism. ADH converts alcohol in the liver to aldehyde, while ALDH converts the aldehyde into acetate and water. Because of the critical physiological roles that these enzymes play in human metabolism, the genetic loci controlling their syntheses and variation should be more highly conserved. Figure 2.7 is a principal component's plot of the first versus the second scaled eigenvectors of an R-matrix consisting of allelic frequencies of three ADH and ALDH loci from eleven world population (Thomasson et al., 2001). The two eigenvectors account for 98% of the total variation. This plot reveals a number of clusters that make no obvious sense when the evolutionary history of these groups is taken into account. For example, American Samoans, Western Samoans, and the Altai form a distinct cluster. Mongols and Hawaiians form another cluster. Judging from the plot of the alleles that contribute to the dispersal of the populations shown in Figure 2.7, the frequencies of the ADH3*2 allele appear to separate the Samoan-Altai cluster from other populations (see Figure 2.8). Frequencies of ADH2*1 and ALDH2*1 are responsible for the Buryat-Kazah cluster. If the common history and ancestry of these groups fails to produce similar allelic frequencies and synthetic clusters, what factors are responsible for creating the observed clusters? Unlike VNTRs or STRs that reflect evolutionary stochastic processes, the genes that encode for physiologically vital enzymes or proteins are most likely effected by natural selection and/or a common history of alcohol consumption. Thus, loci such as hemoglobin, ADH and ALDH should not be utilized to reconstruct human phylogeny.

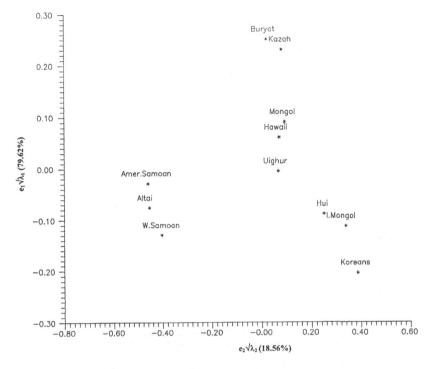

Figure 2.7. Principal components analysis of an R-matrix based on three loci (ADH2, ADH3 and ALDH2) from 11 populations. ADH, alcohol dehydrogenase; ALDH, aldehyde dehydrogenase.

## Interpopulation cluster analysis

A cluster analysis of three Siberian indigenous populations (Nivkhi, Evenki, and Udege) was conducted using mtDNA haplotypes. Kimura's (1980) distance statistic was utilized to measure the genetic distances between mtDNA lineages. The distances were clustered using unweighted pair group method with arithmetic mean (UPGMA) analysis. These three populations were easily separated by the cluster analysis. The Evenki and the Udege tended to cluster together more closely than with the Nivkhi (see Figure 2.9). These clusters represent the known history of the populations, with the Udege being the most easterly extension of the Evenki.

The analysis of molecular variance (AMOVA) was used to partition the genetic variation into components presenting variation between and within populations (Excoffier *et al.*, 1992). AMOVA of mtDNA haplotypes indicates that significant variation ($p < 0.001$) exists among these

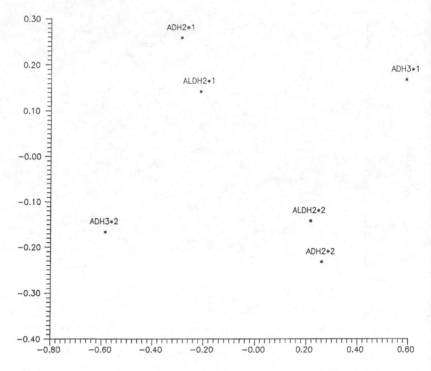

Figure 2.8. A plot of the ADH and ALDH alleles responsible for the dispersion of the populations shown in Figure 2.7.

populations (Excoffier, 1995). In addition, there is significant variation between individuals within populations ($p < 0.001$). However, based on mtDNA, the variation observed among the Evenki brigades is minimal (3.31%) but statistically significant ($p < 0.03$).

### Intrapopulation analyses

Most reconstructions of human phylogeny, using molecular genetic markers, are based on an assumption that the samples are random representations of the village or population. Rarely are adequate genealogical and demographic data collected in order to determine how the genes are distributed within the population. Usually, "get-what-you-can" samples are presented as random, unrelated individuals. Rarely are genealogies of the participants reconstructed in order to determine the nature of the consanguineal or affinal relationships among the members of the sample.

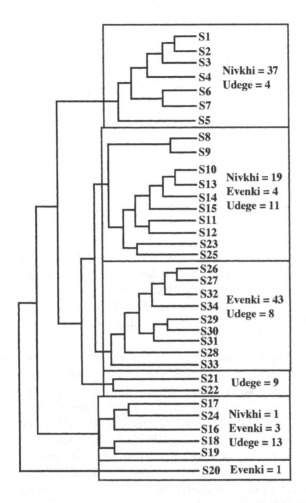

Figure 2.9. Cluster analysis of populations using mitochondrial DNA haplotypes. Kimura's 1980 method was used to compute genetic distances. The unweighted pair group method with arithmetic mean (UPGMA) was utilized for clustering the distances.

In 1996, McComb observed that individuals belonging to Altai clans could on average be correctly classified with 72% accuracy using discriminant function analysis of three VNTR loci. Each individual was entered into a linear discriminant analysis and classified into one of four patrilineal clans, namely Irkit, Todosh, Kipchak, and unassigned (i.e. other clans or the clan affiliation was unknown). Eighty percent of individuals from the Todosh clan were classified into the "True Group", while

40    *M.H. Crawford* et al.

Table 2.11. *Linear discriminant analysis of patrilineal clans of Mindur-Sokhon based on VNTRs*

| Put into group | True Groups | | | |
|---|---|---|---|---|
| | Irkit | Todosh | Kipchak | Unassigned |
| Irkit | 9 | 0 | 1 | 9 |
| Todosh | 2 | 8 | 1 | 5 |
| Kipchak | 1 | 0 | 6 | 8 |
| Unassigned | 0 | 2 | 2 | 9 |
| Total number | 12 | 10 | 10 | 31 |
| % Correct | 75% | 80% | 60% | 29% |

75% of Irkit were correctly classified (see Table 2.11). Persons belonging to the Kipchak patrilineal clan were classified correctly 60%, while only 29% of the unassigned were correctly classified. The expectation of correct assignment based on random factors should be 25% for each of four categories. From these analyses it is clear that members of the same clan share similar or identical genes.

Cluster analysis was conducted using 5 VNTR loci of individuals from Mendur-Sokhon who revealed their clan membership (see Figure 2.10). The individuals of each clan were compared utilizing a genetic similarity matrix. The genetic distance between any two individuals is determined by the number of matching alleles (in this case 0 to 10 alleles). The resulting similarity matrix was clustered using UPGMA cluster analysis (Sokal and Michener, 1958). In most observed clusters, specific clans tend to dominate the analysis. For example, five Irkit clan members cluster together with one Todosh member in the top cluster of the dendrogram. Similarly, three Todosh members form two separate clusters. One cluster consists of representatives from each of the clans. These results suggest that there is an underlying genetic structure within the clans, with some exchanges of mates between the clan groups.

A similar cluster analysis of clans from the Altai, based on nine protein loci (GM, KM, PGM, FA13, FB13, GC, PLG, ACP, and ESD) reveals some grouping (see Figure 2.11). However, a Mantel test indicates that there is no correlation between the VNTR and the protein matrices, which suggests that the two may show the actions of different evolutionary processes. The advantage of this cluster approach over discriminant analysis is that no specific group is specified during the analysis. Thus, this analysis may not be biased by any preconceptions about the underlying structure.

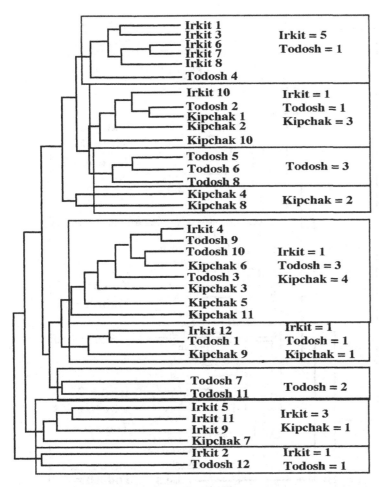

Figure 2.10. Cluster analysis, based on VNTR markers, of individuals belonging to patrilineal clans among the Kizhi-Altai of Mendur-Sokhon.

The final cluster analysis, which included the village of Surinda and its brigades, was performed using mitochondrial DNA haplotypes (Torroni *et al.*, 1993). This analysis relied on Kimura's distance statistic to measure genetic distances between mtDNA lineages. The distances were clustered using UPGMA analysis. Figure 2.12 reveals some patterning, brigade numbers 1 and 6 are separated from the other brigades, but cluster with Surinda. That should not be surprising since Surinda contains Evenki elders who no longer herd reindeer. The parents and the children of many of the reindeer herders reside either permanently or seasonally in the

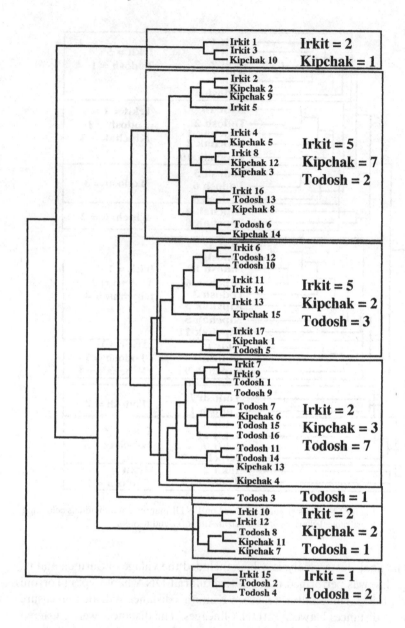

Figure 2.11. Cluster analysis, based on protein markers (GM, KM, PGM, FA13, FB13, GC, PLG, ACP, and ESD) of clan members of Kizhi-Altai.

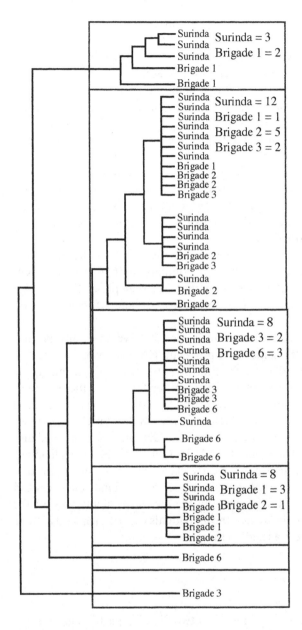

Figure 2.12. Cluster analysis based on mitochondrial DNA (mtDNA) haplotypes for Evenkis from Surinda and its work brigades.

44     *M.H. Crawford* et al.

village, while the adults maintain the herds, and live in tents (*chums*) in the *taiga* throughout the year.

Based upon mtDNA, the Altai populations share the B haplogroup and 9 bp deletion with Mongolia, several Native American populations, and coastal Asia. This distribution of the B haplogroup has prompted several researchers to argue for a Central Asian origin of Native Americans (Merriwether *et al.*, 1995; Sukernik *et al.*, 1996). The Y-chromosome markers support a Central Asian/Mongolian source for one of the migrations into the Americas.

In regards to the examination of the intrapopulation structure of the Evenki village of Surinda and its reindeer-herding brigades, some patterning became apparent. Even with the limited numbers of haplotypes identified in mtDNA, mitochondrial DNA can be used to measure statistically the patterns of gene distribution.

VNTRs appear to provide a better measure of both the inter- and intrapopulation structuring of human populations than does mtDNA. The R-matrix analysis demonstrated that Siberian populations cluster with Native Americans and differ from European and African comparative groups.

Until the 1930s the reindeer were herded by Evenki family groups, usually belonging to the same patrilineal clan. However, with the attempted collectivization of the Evenki tribes by the Soviet government, the ownership of the herds was transferred from the families to the collective. Instead of the family units herding the reindeer, the collectives organized "brigades" consisting of an assortment of related and unrelated individuals. A small sample of brigade members was identified and classified into their true group by discriminant analysis. Only 57% of the individuals entered into the analysis were correctly classified, thus, suggesting that these work brigades do indeed consist of a melange of family members and unrelated individuals. Both of the discriminant function analyses for the Evenki and the Altai indicate that the individuals who participated in these studies do not represent a random sample.

### Conclusion

This chapter summarizes the genetics of two Siberian pastoral groups, the Kizhi-Altai and the Evenki from Surinda and Poligus. The Kizhi-Altai are cattle herders, while the Evenki are reindeer herders. These populations are compared with each other and with surrounding Siberian indigenous populations, plus, for some genetic loci they are compared with a world-wide sample. What is unique about this chapter is: (1) the comparisons are based on a large number of genetic loci that sample many regions of the genome. The following genetic markers have been characterized for these two pastoral populations: standard blood markers (blood groups, serum and erythrocytic proteins), short tandem repeats (STRs), variable number tandem repeats (VNTRs) including DAT1, and COL1A2, Y-chromosome markers, apolipoproteins (APOA4, APOE, APOH), alcohol dehydrogenase (ADH) and aldehyde dehydrogenase (ALDH); (2) in addition to the standard interpopulation analyses, this chapter contains several analyses of gene distributions within the pastoralist populations.

The cluster analyses and the discriminant function analyses used in this chapter demonstrate the importance of considering population subdivision and structuring, even in small, highly isolated populations such as the Evenki and the Kizhi-Altai. The assumption that in these populations genes are randomly distributed is not correct. Patrilineal lineages with clusters of similar genes form the underlying genetic structure of the pastoral populations of Siberia. These lineages make up tribes or settlements, which in turn are part of the specific ethnic groups, i.e., Evenki or Kizhi-Altai. These ethnic groups constitute national *okrugs* or independent countries. Thus, populations of Siberia must be viewed at an appropriate hierarchical level.

On the basis of these genetic data, what can be said about the population structures of the Evenki and the Kizhi-Altai? From all of the genetic data discussed in this chapter, it appears that the Altai pastoralists, because of their position on the crossroads between Asia and Europe, contain more genetic variation than do the Evenki. A plot of the genetic distance from the centroid of distribution versus mean per locus heterozygosity measured with five VNTR loci (McComb *et al.*, 1996) indicates that the Altai has experienced greater gene flow than the Evenki communities. In the principal components plot, Altai also is closer to European populations than are the Evenki populations. Because of their mobility, the Evenki have successfully evaded Russian settlers and administrators and have maintained their genetic uniqueness.

With the exception of ADH and ALDH, most of the genetic markers reveal similar affinities among the populations of Siberia. The ADH and ALDH loci, apparently indicate the action of natural selection among populations that do not have recent common descent. Thus, groups with similar gene frequencies form clusters that are usually interpreted to represent historical connection.

## References

Awadalla, A., Eyre-Walker, M. and Smith, J.M. (1999). Linkage disequilibrium and recombination in Hominid mitochondrial DNA. *Science* **286**, 2524–2525.

Bamshad, M., Watkins, W.S., Dixon, M.E., Jorde, L.B., Rao, B.B., Nadu, J.M., Prasad, B.V., Rasanayagam, A. and Hammer, M.F. (1998). Female gene flow stratifies Hindu castes. *Nature* **395**, 651–652.

Bonne, B., Godberg, M., Ashbel, S., Mourant, A.E. and Tills, D. (1971). South Sinai Beduin. A preliminary report on their inherited blood factors. *American Journal of Physical Anthropology* **34**, 397–408.

Bowles, G.T. (1977). *The People of Asia*. New York: Scribner.

Calderon, R. (2002). Genetic structure of the Basque herders of Northern Spain. In *Human Biology of Pastoral Populations,* ed. W.R. Leonard and M.H. Crawford, pp. 50–63. Cambridge: Cambridge University Press.

Cavalli-Sforza, L.L. and Bodmer, W.F. (1971). *The Genetics of Human Populations*. San Francisco: W.H. Freeman.

Cavalli-Sforza, L.L., Menozzi, P. and Piazza, A. (1994). *The History and Geography of Human Genes*. Princeton, NJ: Princeton University Press.

Crawford, M.H. (1998). *The Origins of Native Americans. Evidence from Anthropological Genetics*. Cambridge: Cambridge University Press.

Crawford, M.H. and Enciso, V.B. (1982). Population structure of circumpolar groups of Siberia, Alaska, Canada, and Greenland. In *Current Developments in Anthropological Genetics,* Vol. 2. *Population Structure and Ecology*, ed. M.H. Crawford and J.H. Mielke, pp. 51–91, New York: Plenum.

Eriksson, A.W. (1973). Genetic polymorphism in Finno-Ugrian populations. *Israel Journal of Medical Sciences* **9**, 1156–1170.

Excoffier, L. (1995). WinAMOVA 3.1 (analysis of molecular variance). Geneva: University of Geneva.

Excoffier, L., Smouse, P. and Quattro, J. (1992). Analysis of molecular variance inferred from metric distance among DNA haplotypes: application to human mitochondrial DNA restriction data. *Genetics* **131**, 479–491.

Forsyth, J. (1992). *A History of the Peoples of Siberia*. Cambridge: Cambridge University Press.

Harding, R.M. (1992). VNTRs in review. *Evolutionary Anthropology* **1**, 62–71.

Harpending, H. and Jenkins, T. (1973). Genetic distances among southern African populations. In *Methods and Theories of Anthropological Genetics,* ed. M.H. Crawford and P.L. Workman, pp. 177–199. Albuquerque, NM: University of New Mexico Press.

Harpending, H. and Ward, R.H. (1982). Chemical systematics and human populations. In *Biological Aspects of Evolutionary Biology*, ed. M.H. Nitecki, pp. 213–256. Chicago: University of Chicago Press.

Jeffreys, A.J., Wilson, V. and Thein, S.L. (1985). Individual-specific 'fingerprints' of human DNA. *Nature* **316**, 76–79.

Jeffreys, A.J., Royles, N.J., Wilson, V. and Wong, Z. (1988). Spontaneous mutation rates to new length alleles at tandem-repetitive hypervariable loci in human DNA. *Nature* **332**, 278–281.

Jorde, L.B. (1980). The genetic structure of subdivided human populations. In *Current Developments in Anthropological Genetics*, Vol. 1. *Theory and Methods*, ed. J.H. Mielke and M.H. Crawford, pp. 135–208. New York: Plenum.

Kamboh, M.I., Crawford, M.H., Aston, C.E. and Leonard, W.R. (1996). Population distributions of APOE, APOH, and APOA4 polymorphisms and their relationships with quantitative plasma lipids levels. *Human Biology* **68**, 231–244.

Karafet, T.M., Zegura, S.L., Posukh, O., Osipova, L., Bergen, A., Long, J., Goldman, D., Klitz, W., Hanihara, S., de Knijff, S., Wiebe, V., Griffiths, R.C., Templeton, A.R. and Hammer, M.F. (1999). Ancestral Asian source(s) of New World Y-chromosome founder haplotypes. *American Journal of Human Genetics* **64**, 817–831.

Kimura, M. (1980). A simple method for estimating evolutionary rates of base substitutions through comparative studies of nucleotide sequences. *Journal of Molecular Evolution* **16**, 111–120.

Lefevre-Witier, P. 1982, Ecology and biological structure of pastoral Isseqqamaren Tuaregs. In *Current Developments in Anthropological Genetics*, Vol. 2, ed. M.H. Crawford and J.H. Mielke, pp. 93–124, New York: Plenum.

Levin, M.G. (1963). *Ethnic Origins of the Peoples of Northeastern Asia*, ed. H.N. Michael. Toronto: University of Toronto Press.

Malecot, G. (1948). *Les Mathematiques de l'Heredite*. Paris: Masson.

McComb, J., Blagitko, N., Comuzzie, A.G., Schanfield, M.S., Sukernik, R.I, Leonard, W.R. and Crawford, M.H. (1995). VNTR DNA variation in Siberian indigenous populations. *Human Biology* **67**, 217–229.

McComb, J., Crawford, M.H., Osipova, L., Karaphet, T., Posukh, O. and Schanfield, M.S. (1996). DNA interpopulational variation in Siberian indigenous populations: The Mountain Altai. *American Journal of Human Biology* **8**, 599–607.

Merriwether, D.A., Rothhammer, F. and Ferrell, R.E. (1995). Distribution of the four founding lineage haplotypes in Native Americans suggests a single wave of migration for the New World. *American Journal of Physical Anthropology* **98**, 411–430.

Miller, R.A. (1991). Genetic connections among the Altaic languages. In *Sprung from Some Common Source: Investigations into the Prehistory of Languages*, ed. S.M. Lamb and E.D. Mitchell, pp. 293–327. Stanford: Stanford University Press.

Mitchell, R.J., Howlett, S., White, N.G., Federle, L., Papiha, S.S., Briceno, I., McComb, J., Schanfield, M.S., Tyler-Smith, C., Osipova, L., Livshits, G. and Crawford, M.H. (1999). Deletion polymorphism in the human COL1A2 gene: genetic evidence of a non-African population whose descendants spread to all continents. *Human Biology* **71**, 901–914.

Mitchell, R.J., Howlett, S., Earl, L., White, N.G., McComb, J., Schanfield, M.S., Briceno, I., Papiha, S.S., Osipova, L., Livshits, G., Leonard, W.R. and Crawford, M.H. (2000). The distribution of the 3′ VNTR polymorphism in the human dopamine transporter gene (DAT1) in world populations. *Human Biology* **72**, 295–304.

48      *M.H. Crawford* et al.

Okladnikov, A.P. (1950). The initial stages of formation of the people of Siberia: the population of the Cis-Baykal region during the Neolithic and Early Bronze ages. *Sovetskaya Etnografiya* 2, Moscow.

Okladnikov, A.P. (1964). Ancient population of Siberia and its culture. In *The Peoples of Siberia*, ed. M.G. Levin and L.P. Potapov), pp. 13–98, Chicago: University of Chicago Press.

Potapov, L.P. (1964). The Altays. In *The Peoples of Siberia*. ed. M.G. Levin and L.P. Potapov, pp. 305–341. Chicago: University of Chicago Press.

Santos, F.R., Pandya, A., Tyler-Smith, C., Pena, S.D.J., Schanfield, M.S., Leonard, W.R., Osipova, L., Crawford, M.H. and Mitchell, R.J. (1999). The Central Siberian origin for Native American Y chromosomes. *American Journal of Human Genetics* 64, 619–628.

Schurr, T.G., Sukernik, R.I., Starikovskaya, Y.B. and Wallace, D.G. (1999). Mitochondrial DNA variation in Koryaks and Itel'men: population replacement in the Okhotsk Sea-Bering Sea region during the Neolithic. *American Journal of Physical Anthropology* 108, 1–39.

Shinka, T., Tomita, K., Toda, T., Kotliarova, S.E., Lee, J., Kuroki, Y., Jin, D.K., Tokunaga, K., Nakamura, H. and Nakahori, Y. (1999). Genetic variations on the Y chromosome in the Japanese population and implications for modern human Y chromosome lineage. *Journal of Human Genetics* 44, 250–255.

Shriver, M.D., Jin, L., Boerwinkle, E., Deka, R., Ferrell, R.E. and Chakraborty, R. (1995). A novel measure of genetic distance for highly polymorphic tandem repeat loci. *Molecular Biology and Evolution* 12, 914–920.

Sokal, R. and Michener, C. (1958). A statistical method for evaluating systematic relationships. *The University of Kansas Science Bulletin* 22 (38,2), 1409–1437.

Sokal, R., Oden, N.L. and Thomson, B.A. (1999). A problem with synthetic maps. *Human Biology* 71, 1–13.

Starikovskaya, Y.B., Sukernik, R.I., Schurr, T.G., Kogelnik, A.M. and Wallace, D.C. (1998). MtDNA diversity in Chukchi and Siberian Eskimos: implications for the genetic history of Ancient Beringia and the peopling of the New World. *American Journal of Human Genetics* 63, 1473–1491.

Sukernik, R.I., Schurr, T.G., Starikovskaya, E.B. and Wallace, D.C. (1996). Mitochondrial DNA variation in Native Siberians with special reference to the evolutionary history of American Indians. I. Studies on restriction polymorphism. *Russian Journal of Genetics* 32, 376–383.

Thomasson, H.R., Crawford, M.H., Zeng, D., Deka, R., Goldman, D., Khartonik, A.M., Mai, K., Ostrovsky, Y.M., Ching, C., Segal, B. and Li, T.-K. (2001). Alcohol and aldehyde dehydrogenase genotypic variation among Chinese, Siberian, North Amerindian populations. *Human Biology* (submitted).

Torroni, A., Sukernik, R.I., Schurr, T.G., Starikovskaya, Y.B., Cabell, M.F., Crawford, M.H., Comuzzie, A.G. and Wallace, D.C. (1993). mtDNA variation of aboriginal Siberians reveals distinct genetic affinities with Native Americans. *American Journal of Human Genetics* 53, 591–608.

Vasilevich, G.M. and Smolyak, A.V. (1964). The Evenks. In *The Peoples of Siberia*, ed. M.G. Levin and L.P. Potapov. Chicago: University of Chicago Press.

Workman, P.L. and Jorde, L.B. (1980). The genetic structure of the Aland Islands. In *Population Structure and Genetic Disease*, ed. A.W. Eriksson, H. Forsius, H.R. Nevalinna and P.L. Workman, pp. 487–508. New York: Academic Press.

Zerjal, T., Dashnyam, B., Pandya, A., Kayser, M., Roewer, L., Santos, F.R., Schiefenhovel, W., Fretwell, N., Jobling, M.A., Harihara, S., Shimizu, K., Semjidmaa, D., Sajantila, A., Salo, P., Crawford, M.H., Ginter, E.K., Evgrafov, O.V. and Tyler-Smith, C. (1997). Genetic relationships of Asians and Northern Europeans, revealed by Y-chromosomal DNA analysis. *American Journal of Human Genetics* **60**, 1174–1183.

# 3 Genetic structure of the Basque herders of northern Spain

ROSARIO CALDERÓN

## The Basques: a general overview

Europe is the mainland area of the Old World whose populations have been and are being most widely analyzed from an anthropological point of view. Nowadays there is a considerable body of data dealing with mating systems, including consanguineous marriages, and genetic diversity patterns of European populations. This fact, associated with the well-documented knowledge of their prehistory, archaeology, paleoanthropology and linguistics have facilitated and accelerated the most modern studies leading to interpretation of the genetic map of Europe and reconstruction of its population history. Analysis of classical genetic markers together with molecular genetic polymorphisms, from the nuclear and mithocondrial DNA genome, are today providing decisive and valuable contributions towards these essential goals of the anthropology.

At present, some interesting anthropological populations are settled in Europe. Cavalli-Sforza *et al.* (1994) selected a group of 42 populations from around the world for genetic studies using geographical, ethnic and linguistic criteria. Of this group, seven were from Europe and included Basques, Sámi (Lapps) and Sardinians together with four other national populations.

It has been shown from a great number of surveys that Basques, a Caucasian population, show notable differences in classical gene frequencies from other European populations. Consanguinity structure has shown appreciable levels of close consanguineous unions (i.e. uncle–niece/aunt–nephew and first cousins) and the temporal trend pattern of inbreeding registered in the Basque area during the past two centuries fits well with the pattern observed in other large European populations (Calderón *et al.*, 1993).

The Basque territory is small, moderately mountainous and situated at the western side of the major mountain range, the Pyrenees, astride the frontier between France and Spain (Figure 3.1). The northern extreme of

50

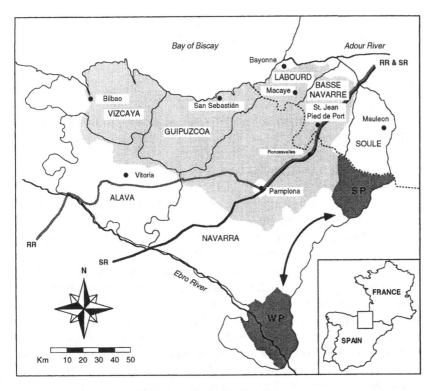

Figure 3.1. Map of the Basque territory with their main territorial administrative divisions in Spain and France. The map also contains: (i) The geography of Basque language or *Euskera* (light-shaded); (ii) The Santiago Route (**SR**) to Compostela as well as the main Roman Road (**RR**) crossing Western Pyrenees. **SR** and **RR** were coincident to Pamplona; and (iii) The longest-range transhumance in Navarre. The transhumance areas are shown as dark-shaded and direction of flocks is indicated by a double-headed arrow. **WP**, winter pastures; **SP**, summer pastures. For further details see text.

the Basque region borders on the Atlantic Ocean which gives it a predominantly moist Atlantic climate. No important geographical barriers exist with neighboring territories and therefore geographical isolation does not seem to have been a critical factor. During the Roman Empire and the Middle Ages the Basque region was a trade corridor between the Iberian Peninsula and the rest of Europe. The famous pilgrimage route to Santiago de Compostela and an important Roman Road went through the Basque region via the Navarre and Alava territories.

The Basque lands in Spain are governed by the Basque Country Autonomous Community (BCAC) ($10\,421\,\text{km}^2$) and the Navarre Chartered

Community (NCC) (7261 km$^2$), which are administratively unconnected. The BCAC comprises three provinces: Guipúzcoa, Vizcaya and Alava whose provincial capitals are San Sebastián, Bilbao and Vitoria, respect- ively. The capital of Navarre is Pamplona. The French Basque Country (FBC) (3065 km$^2$) is divided into the regions of Labourd, Basse Navarre and Soule. All of these territories make up the whole "Basque area" which covers in all 20747 km$^2$ and had 2808390 inhabitants at the time of the 1975 census. The BCAC is known in Spain as Basque Country, thus we shall use both of these terms as synonymous.

The Basque language, or *Euskera*, is not Indo-European and is notice- ably different from languages spoken by neighboring populations. Basque language, often classified as a linguistic isolate, has been clustered recently with the Abkhaz, North-Caucasians, Sino-Tibetans and Na-Dene lan- guages (Chen et al., 1995). The geographical area of the Basque language does not coincide with the limits of the Basque area as a whole as this has changed over time; it also exhibits geographical dialectical differences. The diachronic geography of the Basque language started to be clearly revealed from the sixteenth century on and only the province of Guipúzcoa has remained relatively unchanged on the Basque linguistic map. The Basque territories where the *euskera* recession has been greatest, are those of Alava and Navarre. The term "Euskera area" will be used here to mean that part of the Basque area whose inhabitants have spoken predominantly Basque language over centuries The present day Euskera area covers around 9300 km$^2$ of which 2800 km² are in French territory and 6500 km² in Spanish territory (4100 km² in the Basque Country and 2400 km² in Navarre, mostly in its northern part).

Two centuries ago the main Basque cities were quite small and Basques as a whole were a traditional rural society. At present, the population settled into the Basque area constitutes a modern industrial society with large and important towns causing its demographic density to be appreci- ably higher than that of surrounding populations. Basques have under- gone the different stages of European technological development at similar periods to other European populations.

The population of the Basque area has been unevenly distributed and this might be a logical outcome of the early industrial development of the Basque Country by the end of nineteenth century. A large number of immigrants from both surrounding and distant Spanish provinces were attracted by this industrialization and settled in the initially small cities (e.g., San Sebastián, Bilbao and Vitoria each contained less than 10000 people in the nineteenth century). Outside urban centres, people have traditionally lived in nuclear villages and in scattered rural households (*caseríos*). Dispersed settlement population patterns are common all over northern Spain.

Nowadays the demographic size of BCAC numbers 2 107 307 inhabitants (Census of 1996, EUSTAT 1998) and Vizcaya province accounts for 54% of this number. The present-day metropolitan area of Bilbao, comprising the city of Bilbao itself and its surrounding industrial belt, holds over a million people. This conurbation accounts for a third of the Basque area's total population. From 1830 until the present the population density of the BCAC has fluctuated between twice and four times that of Spain in general.

## Traditional economic activities in the Basque area

### Pastoralism and its biocultural effects

Although the Basque area underwent early industrial growth towards the end of the nineteenth century, a traditional economy based on small-scale agriculture is still an efficient and extended productive sector. Fishing and sheep-herding are also other important ways of life there. Pastoralism continues as a deep-rooted activity, by global standards, Basque shepherds can be deemed experts.

Pastoralism in the Basque area was and indeed still is basically concerned with sheep-herding. Pastoralism should not be thought of as a pre-Neolithic human activity. No evidence for domesticated animals, except for dogs, exists in the Basque area before around 5000 BP (i.e., around the time when the construction of megalithic monuments began). In the burial found in the cave of Marizulo (Guipúzcoa) a human skeleton was accompanied by sheep and dog bones. Radiocarbon dating has placed this context at 5285±65 BP (Arias 1994*a*, cited in Zilhao 1995).

The Basque anthropologist Barandiarán established a close relationship between Megalithic cultures (i.e., dolmens) all over the Basque territory and pastoral activities. Montane shepherding areas and also important megalithic trackways are pockmarked with these prehistoric monuments.

The wet climate of the Atlantic-facing slopes of the Basque area is a major conditioning factor for pastoral life there. There is an autochthonous Basque breed of sheep, the *latxa*, which is highly prized for the excellence of its milk, though its wool and meat are of poor quality. This sheep adapts well to the wet Atlantic climate.

In the fourteenth century, the Castilian Crown created in Spain an organization, named the *Mesta*, to bring together the existing associations of sheep-owners. The *Mesta* was entrusted with the supervision and control of the complex systems whereby the great migratory flocks – *transhumance* – were moved across Spain.

Under the control of the *Mesta*, the wool industry brought the Castilians into closer contact with Flanders (present-day Belgium), which was then the most important wool market in Europe. The growth of the *Mesta* and the expansion of the wool trade with northern Europe stimulated the development of the ports of northern Spain – San Sebastián and Santander – among others (Elliott, 1963, p. 33). The flocks of the *Mesta* reached their peak in the mid 1520s, at around three and a half million head of sheep.

Official statistics dealing with agropastoral activities as well as Basque shepherd demography are available from Basque Government publications. Traditionally, Basque shepherding lifestyle has usually been analyzed from different standpoints, but special attention has been given to pastoral mobility and its effects on families. Pastoralism in the Basque territory mainly concerns with short-range mobility and it has been and still is practiced in Guipúzcoa, Vizcaya, northern Alava and northwestern Navarre. The only important Basque sheep transhumant movements involving long-range geographical distances (i.e. around 140 km on average) is that from eastern Navarre, between the Pyrenean valleys of Roncal and Salazar and southern Navarre and is known as "the winter transhumance" (see Figure 3.1). There the broad heathlands of "Bárdenas Reales" and other surrounding areas beside the Ebro River provide good pasturing lands; this has caused numerous shepherds with their flocks to spend the bulk of their time (i.e. the whole winter) in this geographical area far away from their families.

This kind of pastoralism seems have had linguistic rather than genetic consequences. Basque-speaking Pyrenean shepherds had contact with the Spanish-speaking population of southern Navarre during their long and repeated winter stays. When they returned home, for the rest of the year, they taught their recently learned Spanish to their own families and it marked a progressive and inevitable loss of the Basque language in those Navarrese Pyrenean valleys. The importance of Spanish in northern Navarre was noted from the nineteenth century onwards.

We have not detected in the scientific literature any population studies dealing with genetic characteristics of Basque shepherd communities. However, data from marital behaviour of Pyrenean Navarrese populations reveal that it closely resembles that shown by the rest of Basque rural population from Spain (Toja, 1987). People tend to marry in the same place as where they were born, in ways which are hardly distinguishable from those of their farmer neighbors. Premarital mobility of consanguineous marriages is very small and the amount of close consanguinity (uncle–niece, aunt–nephew and first cousin marriages) is not so high when compared with the high frequencies of recorded second cousin matings (Zudaire, 1984). Consequently, the Basque shepherd community was neither mobile enough nor demographically important enough to serve as a genetic homogenizing mechanism for the Basque population.

### Genetic characteristics of the Basque population

Nowadays the demographic size of the total human population is far higher than it was historically. At present nearly 6000 million people live in the world and it has been estimated that there are about 5000 populations based on the different number of languages spoken today (Cavalli-Sforza *et al.*, 1988). One of these populations possessing an ancestral language of their own is the Basques.

Studies dealing with genetic structure analysis from many hundreds of human populations around the world, have shown that particular frequencies for certain classical alleles appear confined in major continental groups and also may be common enough in some particular geographical regions or in specific populations. The rhesus haplotype *Ro (cDe)* and the genetic variant of the Duffy locus, the silent *Fy* allele, present distinctive frequencies in Africans. The *GM*3 5** haplotype and the *HLA-A2* allele are considered as excellent anthropogenetic markers in Europeans. The *HLA-Cw4* in Mediterranean populations and the rhesus-negative and B blood group gene show frequencies in Basques. Frequencies of these alleles may be specially informative for gene admixture estimates between populations.

The genetic characterization of the Basque population has been described for almost all traditional genetic polymorphisms and also inter-variability of Basque populations has been recently examined by using classical gene frequencies. These studies have concluded that there is significant heterogeneity within the Basque population and integration of data from linguistic, historical and demographic studies has been required to explain these genetic diversity patterns. An example will be provided here from GM and KM immunoglobin (Ig) allotype frequencies in Basques from Spain and France; sources of data together with some approaches on the origin of the Basque population can be found elsewhere (Dugoujon *et al.*, 1989; Hazout *et al.*, 1991; Calderón *et al.*, 1998). For these purposes the first study consisted of analyzing blood samples of French Basques from Macaye (Labourd), Saint Jean Pied de Port (Basse

Navarre) and Mauleon (Soule); the latter survey has recently analyzed Basques from Alava, Guipuzcoa and Vizcaya provinces in Spain. Other complementary genetic information, which extends the knowledge of the Basque population for GM and KM markers to Basques from northern Navarre (Spain) (Calderón et al., 2000), will be used here for discussion purposes. The seven Basque study subpopulations analyzed correspond to the major administrative divisions of Basque territory as a whole.

It is well known that observed patterns of genetic variation give insight into population similarities. These relationships have been shown by using distance-based methods which transform original data on gene frequencies into a single measure of evolutionary distances. These analyses usually end up reconstructing the phylogeny by means of the application of different algorithms (e.g. unweighted pair group method with arithmetic mean, or UPGMA, neighbor-joining) which take into account times of divergence between populations or individuals.

The most frequent GM haplotypes in Basques are those common in Europeans GM*3 23 5*, GM*3 23' 5*, GM*1,17 23' 21,28, and GM*1,2,17 23' 21,28. (see Calderón et al., 2000, table 3).The main Caucasian haplotype, GM*3 5* (which includes GM*3 23 5* and GM*3 23' 5*) shows low to moderate frequencies in Basques when compared with European values. The lowest Basque frequency of this haplotype has been detected in Basques from Guipúzcoa which also shows the highest frequency of GM*1,17 23' 21,28 with respect to European populations.

The African haplotype GM*1,17 23' 5* (Excoffier et al., 1991) reaches frequencies ranging from 0.02 to 0.01, in most Basque subpopulations. This haplotype has been observed in other populations living in the Iberian Peninsula as well as in Mediterranean Europe. The haplotype GM*1,17 23' 10,11,13,15,16, which is considered a genetic marker of North Asian populations (Matsumoto, 1988), presents polymorphic frequencies in Basques from Guipúzcoa and Vizcaya, while lower values are observed in the other Basque subpopulations.

The rare haplotype GM*1,17 23 21,28 was found to be in polymorphic frequencies across three Basque subpopulations: Alava, northern Navarre and St Jean Pied de Port. This haplotype is missing in the other four Basque subpopulations.

The spatial pattern of the immunoglobin allotypes in the Basque study subpopulations indicates that their genetic isolation from neighboring populations has been broken differentially, the core of the Basque area being the most resistant part to admixture. However, Basques as a whole have maintained their genetic distinctiveness until the present. Guipúzcoa

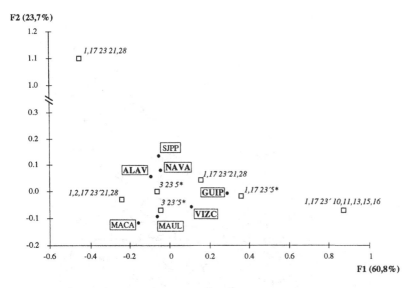

Figure 3.2. Correspondence multivariate analysis based on the *GM* haplotypes in Basque subpopulations from Spain and France. GUI, Guipuzcoa; VIZ, Vizcaya; ALAV, Alava; NAVA, northern Navarre; MACA, Macaye; SJJP, St Jean Pied de Port; MAUL, Mauleon. Basque subpopulations from Spain are in bold. (After Calderón *et al.*, 2000.)

presents the highest genetic distance value with all the Basque subpopulations except with Vizcaya. The non-metric multidimensional scaling performed on the overall Basque population showed that Guipúzcoa is clearly distinguished from the rest of Basque subpopulations, Vizcaya presents an intermediate position and Alava, northern Navarre and St Jean Pied de Port are close together (see Figure 3.2). These observations suggest that (a) Basque geographical distances seem not to be a good predictor of their genetic distances, and (b) although the Basque communities have been defined using the same autochthonous criterion, they seem to show different levels of admixture of a founder Basque population with others. The highest levels of admixture would be present in Alava and the lowest in Guipúzcoa, which would be in good agreement with the Basque linguistic map.

A comparison of Basques with other European populations using the three most representative *GM* European haplotypes, *GM\*3 5\**, *GM\*1,17 23′ 21,28* and *GM\*1,2,17 23′ 21,28* shows that Basques from France are more similar to Europeans than Basques from Spain. (see Figure 3.3).

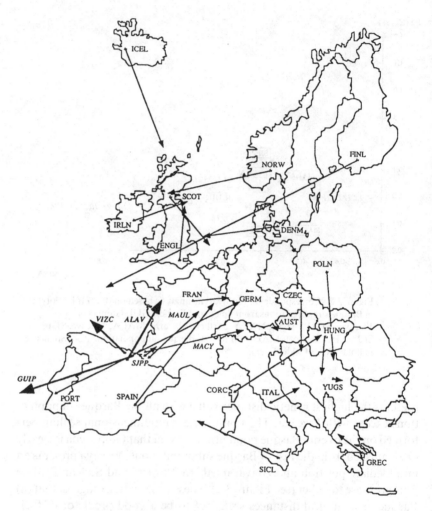

Figure 3.3. Relationships between geographical and genetic patterns. The genetic topology of Europe has been built up using the three most representative GM haplotypes in Caucasians (*GM\*3 5\**; *GM\*1,17 21,28* and *GM\*1, 2,17 21,28*). (After Calderón *et al.*, 1998.)

## Basque population evolutionary history

Defining the characteristics of a population consists in tracing the way it came about, so that it can be shown on what basis and in what forms that process has taken place. Populations arise at a certain time within a geographical area and will remain the same over a long period of time, either in that same geographical area or moving to another. That population will continue to be recognizable for its genetic characteristics and likely retaining genetic exchanges with neighboring populations, and even with geographically distant populations. The place of a population emergence does not necessarily coincide with the current territory of settlement nor even one that might be referred to in historical sources. Movements or fissions of populations mainly from the transition to the Neolithic period should not be disregarded.

The population emergence is the result of a splitting followed by a population growth of that differentiated (segregated) part. This process is associated with the *Founder effect*, in which a relatively small group of people ultimately became the progenitors of a new population who may not necessarily represent the whole range of variation of the ancestral gene pool from the base population. Such a population will try to maintain itself by means of successive expansions that avoid its genetic dilution. The chance of detecting the moment of populational emergence is low, since in some cases there would not be genetic traces. Nevertheless certain moments can be rejected and these would be based on population sizes, human mobilities and mating systems necessary for population emergence to take place.

Basques are a Caucasian population but genetically and linguistically distinct from many other European populations. So they are a paradigm of the complexity of human evolution. The most common argument to explain the Basque peculiarity states that they are a relic population of the Paleolithic humans from Europe, being the result of *in situ* differentiation by genetic drift and selection over a very long time, while they were exceptionally isolated from other populations. However, the proponents of the relic hypothesis have established neither the time and the territory of Basque origin nor the prehistoric and historical changes that Basques have undergone. Because of the lack of ecological niches or important geographical barriers in the Basque area, the stated isolation mechanism is said to be cultural, mainly due to their odd language. The relic hypothesis implicitly requires that Basques have passed through many evolutionary events, in a way quite different from that observed in many other human populations. Because a series of events, each one quite improbable, needs to occur in the relic scenario, it is unlikely that this hypothesis might be valid.

We have proposed elsewhere (Calderón *et al.*, 1998) that the Basque singularity could be plausibly explained by a replacement hypothesis, which states that hunter-gatherers wandering the Basque territory could be

60    R. Calderón

well replaced by a small Neolithic population that occupied that region and introduced new agricultural technology there. These farmers could have been a migrating group coming from a distant place. The Neolithic population could have experienced rapid growth over a few generations, while not mating to any appreciable extent with the neighboring hunter-gatherers. This process would consolidate the genetic characteristics of the newcomers, who soon reached a sufficient population size to resist permanently any possible admixture with other populations (Calderón, 1994). However, some admixture has inevitably occurred, and it has been spatially structured, as their modern gene frequencies show. The smallness of the Basque territory and the low densities of pre-Neolithic populations, (on the order of 0.1 inhabitants per square kilometer) result in a quite small size of this pre-Neolithic group, easily allowing for its replacement. The time of the migration of the farmer group to the Basque region can be estimated at 5000–5500 BP. This date allows us to quantify roughly the number of generations of Basque evolution in Basque territory in order to evaluate the weight of the different evolutionary processes on this population. Migration may have been the result of the pressure of an increase in population size in the native land, or of withdrawal following the conquest of that land by groups of early horsemen who imposed their rule upon farmers (Diamond, 1991).

As has been indicated earlier, *Euskera* shows the closest similarity with the family of the northwestern Caucasian languages (Starotsin, 1989, cited by Renfrew, 1992; Chen *et al.*, 1995) and we have suggested that the Neolithic migrating farmers, who settled in the Basque area, could have been a group of North Caucasian natives. The similarity between *Euskera* and Caucasian languages was noted many years ago (Urreiztieta-Rivera, 1980).

Alleles that are observed in a population can be common or rare and this distinction can be based on the incidence with those that are detected in the population. "Rare" alleles do not imply that they are very rare variant alleles, and would be those whose frequencies are near to polymorphic level (ranging between 0.005 and 0.02), while "common" alleles appear with frequencies higher than 0.02. This categorization is conventional since other similar threshold values could be chosen. Furthermore alleles can also be denominated "rare" for an unexpected presence in a specific population, although these alleles may be otherwise common in other populations or in major continental groups to which that population could be observed as belonging. In any event to establish the distinction between rare and common alleles makes it necessary to study their genetic geography in many populations. Rare alleles may be especially informative in detecting crucial aspects of the evolutionary history of populations. Human Ig allotypes have proven to be highly informative ethnic markers and hence have shown their power as a genetic tool for unveiling complex evolutionary processes. Because of the presence of *GM\*1,17 23′ 10,11,13,15,16* in Basques, we have postulated an ancient admixture of the North Caucasian population with a North Asian one. The highest frequency of this haplotype in Asia has been observed in the Lake Baikal region as well as a cline in all directions (Matsumoto, 1988). The ancient North Asian movement towards the Caucasus could be simultaneous with, but opposite in direction to the Na-Dene migration to America, dated from 10 000 to 14 000 BP. Both migrations might have affected the same population and might be due to the same causes. The low frequencies of *GM\*3 5\** and the high values of *GM\*1,17 23′ 21,28* in Basques, compared with European values, would support the hypothesis that they are a Caucasian population with a low admixture of North Asians.

The polymorphic frequencies of *GM\*1,17 23′ 5\**, the main African haplotype found in Basques, suggest that this population has not been so isolated, as it is commonly argued, from other populations, in particular North African ones, which inhabited the Iberian Peninsula for long periods of history. However this haplotype is also in rather high frequency in many other populations; the most noteworthy appears to be Central Asia where it co-occurs with the presence of *GM\*1,17 23′ 10,11,13,15,16* (Schanfield and Kirk, 1981; Schanfield *et al.*, 1990). Therefore, it has multiple possible origins and Central Asia is an area that probably served as an ancient homeland for both European and Asiatic populations. The Basque frequencies might alternatively be due to the above-mentioned North Asian groups.

When multivariate and phylogenetic methods are applied to GM data in Basques, the northern Navarre subpopulation appears centrally forming a cluster with Alava and St Jean Pied de Port and the rare *GM\*1,17 23 21,28* haplotype contributes most to this clusterization. This genetic pattern could be caused by the main corridor, embracing the Roman and Santiago routes, which connected through Antiquity and Middle Ages until the present all three of these Basque territories (see Figure 3.1). We have hypothesized elsewhere that the occurrence of that rare *GM* haplotype in Basques could be caused by Roman settlers at the start of the Christian Era. Further interpretations can be found in Calderón *et al.* (2000).

We have seen that analysis of geographical genetic variation patterns based on Ig allotypes in Basques has revealed genetic heterogeneity among Basque subpopulations. In the future it would be most desirable to extend, to other genetic markers, this sort of subpopulational analysis, which

62    R. Calderón

reflects a compromise between the proposal stated by L.L. Cavalli-Sforza of taking only one blood sample per population and that from A.C. Wilson of collecting human population blood samples along a geographical grid at more or less evenly spaced locations (Roberts, 1992), but strengthening the anthropological knowledge of the study population from a linguistic, demographic, historical and technological perspective. So, the hypothesis about the origin of Basques as a Neolithic replacement followed with subsequent selective admixtures could be more firmly validated, or eventually rejected.

## References

Calderón, R. (1994). Population structure studies in Europe: the Basques. Lecture at the American Association of Anthropology and Genetics. Annual Meeting. Denver, CO.

Calderón, R., Peña, J.A., Morales, B. and Guevara, J.I. (1993). Inbreeding patterns in the Basque Country (Alava Province, 1831–1980). *Human Biology* 65, 743–770.

Calderón, R., Vidales, C., Peña, J.A., Perez-Miranda, A. and Dugoujon, J.M. (1998). Immunoglobulin allotypes (GM and KM) in Basques from Spain. Approach to the origin of the Basque population. *Human Biology* 70, 667–698.

Calderón, R., Perez-Miranda, A., Peña, J.A., Vidales., C., Aresti, U. and Dugoujon, J.M. (2000). The genetic position of the autochthonous subpopulation of northern Navarre (Spain) in relation to other Basque subpopulations. A study based on GM and KM immunoglobulin allotypes. *Human Biology* 72, 619–640.

Cavalli-Sforza, L.L, Piazza, A. and Menozzi, P. (1988). Reconstruction of human evolution: bringing together genetic, archeological, and linguistic data. *Proceedings of the National Academy of Sciences* 85, 6002–6006.

Cavalli-Sforza, L.L., Menozzi, P. and Piazza, A. (1994). *The History and Geography of Human Genes*. Princeton, NJ: Princeton University Press.

Chen, J., Sokal, R.R. and Ruhlen, M. (1995). Worldwide analysis of genetic and linguistic relationships of human populations. *Human Biology* 67, 595–612.

Diamond, J.M. (1991). The earliest horsemen. *Nature* 350, 250–251.

Dugoujon, J.M., Clayton, J., Sevin, A., Constans, J., Loirat, F. and Hazout, S. (1989). Immunoglobulin (Gm and Km) allotypes in some Pyrenean populations in France. *Collegium Antropologicum* 13, 43–50.

Elliot, J.H. (1963). *Imperial Spain (1469–1716)*. London: Penguin Books.

Excoffier, L., Harding, R.M., Sokal, R.R., Pellegrini, B. and Sanchez-Mazas, A. (1991). Spatial differentiation of Rh and Gm haplotype frequencies in sub-Saharan Africa and its relation to linguistic affinities. *Human Biology* 63, 273–307.

EUSTAT. (1998). *Movimiento Natural de Población*. Vitoria, Spain: Instituto Vasco de Estadística.

Hazout, S., Dugoujon, J.M., Loirat, F. and Constans, J. (1991). Genetic similarity maps and immunoglobulin allotypes of eleven populations from Pyrenees (France). *Annals of Human Genetics* **55**, 161–174.

Matsumoto, H. (1988). Characteristics of Mongoloid populations based on the human immunoglobulins allotypes. *Anthropologischer Anzeiger* **46**, 119–127.

Roberts, L. (1992). How to sample the world's genetics diversity. *Science* **257**, 1204–1205.

Schanfield, M. and Kirk, R.L. (1981). Further studies on the immunoglobulin allotypes (Gm, Am and Km) in India. *Acta Anthropogenetica* **5**, 1–21.

Schanfield, M.S., Crawford, M.H., Dossetor, J.B. and Gershowitz, H. (1990). Immunoglubulin allotypes in several American Eskimo Populations. *Human Biology* **62**, 773–789.

Starotsin, S.A. (1989). Nostratic and Sino-Caucasian. In *Exploration in Language Macrofamilies*, ed. V. Shevoroshkin, pp. 42–66. Bochum: Studienverlag Dr N. Brockmeyer.

Toja, D. (1987). Estructura matrimonial de las poblaciones de dos valles Pirenaicos. Ph.D. thesis. Universidad de Barcelona, Spain.

Urreiztieta-Rivera, I. (1980). Basque and Caucasian: a survey of the methods used in establishing ancient genetic affiliations. Ph.D. thesis. The University of Arizona. University Microfilms, Ann Arbor, MI.

Zilhao, J. (1995). From the Mesolithic to the Neolithic in the Iberian Peninsula. In *The Transition to Agriculture in Prehistoric Europe*, ed. T. Douglas Price. 60th Annual Meeting of American Archeology, USA (May 1995).

Zudaire, C. (1984). Consanguinidad en los valles del Roncal, Aezcoa y Salazar. *Cuadernos de Etnología y Etnografía de Navarra*, 107–127.

# 4 History, demography, marital patterns and immigration rate in South Sinai Bedouins: their effect on the coefficient of inbreeding (F)

E. KOBYLIANSKY AND I. HERSHKOVITZ

Most populations in the world today reside in more or less man-made environments. Relatively few have remained in the natural ambiance in which they had developed for hundreds or even thousands of years, continuing to maintain direct ties with it. One of the latter is the South Sinai Bedouin group, which has adapted biologically, socially and behaviorally to the extreme arid environments of the Middle East. Throughout the South Sinai Peninsula are scattered 10 Bedouin tribes totaling some 10 500 individuals (see Figure 4.1).

Most of the tribes originally came from the Arabian Peninsula, although a few are linked traditionally to Egypt and North Africa, and one derives partly from the southeastern part of the European continent. All the Bedouins of South Sinai are Moslem and make a living primarily from labor outside their territorial base in industrial areas in Israel and Egypt, as well as from raising sheep and camels, fishing, smuggling, and engaging in a primitive type of agriculture.

The main features of the Bedouins in South Sinai are:

(a) Isolation. Topographic barriers, such as the Tih escarpment in the north, the Gulf of Eilat in the east, the Gulf of Suez in the west, and the Red Sea in the south, coupled with efforts of the tribes themselves to retain their cultural uniqueness, have ensured virtual biological isolation of the tribes from the surrounding societies.

(b) Mating patterns. Consanguineous marriages, especially between first cousins are preferred within the Bedouin population of South Sinai. Thus, in decreasing order of preference in selection of a mate is the extended family (first cousins) or Hams (blood feud group), clan, or blocks of extended families (social) units among Bedouins which have no precise anthropological term, among settled Arabs they are known as

64

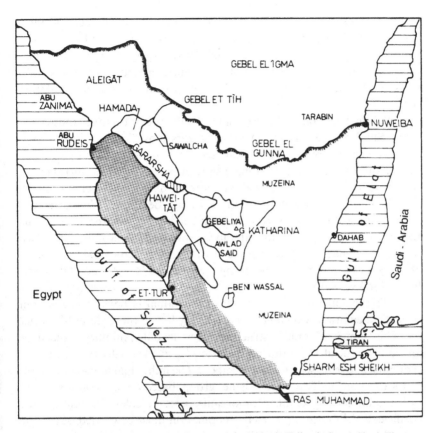

Figure 4.1. Territorial distribution of the Bedouin Tribes in South Sinai. The stippled area demarcates the most arid zone in the Sinai, not claimed by any tribe.

"khamulas", subtribe and tribe. Marriages outside the tribal frame are very limited, as will be noted subsequently.

(c) Uniform environment. The Bedouin residents of South Sinai live in an environment which is uniform in many respects, to wit: an arid climate, an economy primarily based on "outside" labor and traditional occupations, as noted earlier, education which comprises, at best, two or three years of elementary schooling; health care which includes medical services provided from 1967 to 1982 by the Israeli civil administration, prior to which there was reliance only on traditional tribal medical practices. The Bedouin population in South Sinai lacks any social or religious stratifications thus being largely egalitarian.

(d) Population size. Most of the tribes in South Sinai number between

500 and 1000 members, thus comprising relatively small social groups. Prior to the Israeli presence in 1967, the various tribal infrastructures tended to be even smaller, being delimited largely by the nature and "carrying capacity" of the area. Improvement in socioeconomic conditions and health status generally after 1967 led to increase in population size.

(e) Political instability. In the first half of the present century the area of South Sinai passed through many conquests – by Turks, British, Egyptians, Israelis, Egyptians again, Israelis again, and most recently Egyptians for the third time. The penultimate shift of power to Israeli jurisdiction, commencing in 1967, brought about large changes in the social, medical, economic and cultural infrastructures of the Bedouins. This latter upheaval had a direct as well as an indirect bearing on the lives of the Bedouins, initiating a shift to better living standards heretofore not known to them. In many respects the new standards reshaped Bedouin ways of living, thinking, and working (Marx, 1974; Perevolotzky and Perevolotzky, 1979).

(f) Tribal population. The Bedouin individual is born into a certain tribal framework to which he remains firmly linked throughout his or her life. This framework or infrastructure, albeit devoid of military, judicial, or any other official or authoritarian backing, provides for the personal, economic and legal safety of its members. The individual member, in turn, bears a responsibility towards his/her group, based on an "honor system" rather than on any codified or organized set of rules. Calculating the coefficient of inbreeding in such a population must be carried out in the following steps: (1) revealing the history of the tribes, their geographical origin and ethnicity; (2) establishing the biological relationships between the tribes; (3) reconstruction of marriage pattern and migration behavior; (4) obtaining detailed demographic information regarding size of the tribes and families, life pyramid, and polygamy.

## Origin and relationships between South Sinai Bedouin tribes: narrative evidence

The history of the South Sinai Bedouin tribes has yet to be fully clarified. The most reliable sources are probably the weekly reports of the Head of the Santa Katharina monastery in South Sinai. These reports, which have been written virtually ever since the foundation of the monastery at the end of the sixth century, have never been made available for scientific scrutiny, owing to opposition of the local monks. Consequently the main source of information on the history of the South Sinai tribes has perforce been the accounts of ethnographers and anthropologists primarily in the previous century (Burckhardt, 1822; Robinson, 1841; Stanley, 1864), as well as "tales" related by the Bedouins themselves, especially those by Sheikh Muhamad Mardi Abu-Le'ham of the Gebeliya tribe (personal communication).

The southern Sinai Peninsula, although a desert area with harsh topographic conditions was inhabited by human populations even in early prehistoric time, according to the archaeological evidence (Bar-Yosef, 1980). Thus, large settlement sites were found in South Sinai in the Neolithic era (8000–5500 BCE; Bar-Yosef, 1981), through the Chalcolithic era (4000–3150 BCE; Goren, 1980), as well as in the early Bronze Age (3150–2200 BCE; Beit-Aryeh, 1980), the Byzantine period (324–640 CE; Tsafrir, 1970; Finkelstein, 1980), and the early Arab period (640–1291 CE; A. Goren, personal communication). Since the Moslem conquest of Sinai (621 CE), Bedouin tribes from the Arabian Peninsula began a constant infiltration into South Sinai and in the course of time they either displaced or absorbed the native pagan populations known in the literature by the name "Saracens" (Tsafrir, 1970). In view of the fact that ingress or egress of tribes into or out of South Sinai was a relatively rapid process, and the written testimonies date mainly from relatively recent times, it is uncertain which were the first Bedouin tribes to settle in South Sinai after the Moslem conquest and why they disappeared subsequently (Levi, 1987). Ben-David (1978) notes that one of the first tribes to settle in the region was the "Beni-Suleiman" tribe that achieved ascendancy about 700 years ago. Robinson (1841) also mentions this tribe as one of the early ones in the region and conjectures that it should probably be regarded as a relic of the ancient Saracen population. Burckhardt (1822) notes (p. 559) that this tribe settled in all of South Sinai following the first Moslem conquests in 638 CE but was decimated by the Sawalcha and Aleigat Bedouin tribes which invaded South Sinai from Egypt.

The tribes presently found in South Sinai are the Gebeliya, Sawalcha, Hamada, Aleigat, Beni-Wassal, Muzeina, Haweitat, Gararsha, Awlad Said and Ahali et-Tur. Among the first to arrive were probably the Hamada and the Beni-Wassal tribes (Arensburg *et al.*, 1979). Some consider the Beni-Wassal tribe an offshoot of the Hamada tribe, which split off as a consequence of an intratribal clash (Ben-David, 1978). The Bedouin version of the latter has it that the Beni-Wassal clan or khamula (number of extended families that share a common ancestor) joined forces with the

Sawalcha tribe in a battle against the Aleigat tribe, but after the Sawalcha were defeated, an armistice agreement included a clause demanding the expulsion of the Beni-Wassal from the mother tribe (Hamada) and their transfer to the territory of the Muzeina tribe, an ally of the Aleigat in the aforementioned battle.

There is evidence in South Sinai (e.g., the graves of sheikhs, palm tracts) which clearly indicates that the Beni-Wassal tribe was in the not very distant past one of the largest and strongest in South Sinai (Arensburg *et al.*, 1979), whereas the Hamada tribe was always a small group adhering to fixed habitats and apparently never departing from them.

The Hamada tribe is probably the most ancient of the current South Sinai tribes. At the end of the fourteenth century there was increased migration of Bedouin tribes into South Sinai. The first to arrive then was the Aleigat tribe. According to Ben-David (1978), this tribe came from Ullah, a region and city to the northwest of Khedjaz (Saudi Arabia). The penetration of new tribes into South Sinai that began in the thirteenth century provoked resistance from the "veteran" tribes in the regions especially the largest of these, namely, the Sawalcha tribe.

The second wave of migration of Bedouin tribes into South Sinai commenced apparently at the beginning of the fifteenth century. The "veteran" tribes (e.g., Hamada, Beni-Wassal, Nefeat and Ayede) loosened their hold on the southern part of the Sinai Peninsula, most of their members migrating into North Sinai and Egypt; an example are remnants of the Ayede tribe which are currently concentrated in the region of Port Toufik in Egypt. It is very likely that the departure of these tribes was neither rapid nor a single occurrence. Extended periods of drought may have led to occasional massive emigration from the Sinai, but infiltration of individuals from South Sinai into Egypt must have been a long and continuous process. We know that the Egyptians tried to prevent this process and that one of the major problems facing the first and succeeding Pharaohs was to contain this infiltration of desert dwellers into their settled country, but to no avail (Sharon, 1977). At that time, every piece of land vacated in South Sinai was immediately occupied by members of new tribes, so that there was no return of those which had left the territory. The Muzeina tribe, since the ninteenth century, has become one of the most successful in South Sinai. Within a brief span of time it dominated all of southeastern Sinai, from Sharm-el-Sheikh in the south to Nuweiba in the north. Other tribes infiltrated the region as well as the Muzeina. Thus, the Awlad Said tribe occupied the eastern and southern borders of the high mountain. The hold of this tribe in South Sinai was encouraged by the monks at the Santa Katharina monastery who allowed its members to cultivate the remote orchards of the monastery, particularly those in Wadi Isla, and to protect the Dir-Antush monastery at the foot of Gebel Umm-Gumar, which was strategically situated to secure the supply and pilgrimage routes between Et-Tur and the mother monastery. Evidence regarding the geographical origin of the various clans (khamulas, groups of extended families) points to the fact that this tribe comprises a sort of conglomerate of families that had reached South Sinai at different times and had become organized into a new tribal infrastructure. Burckhardt (1822) and Robinson (1841) regarded this tribe as one of the clans of the Sawalcha tribe. Another tribe settling in South Sinai at that time (fourteenth century) was the Gararsha.

The last tribe to settle in South Sinai was the Haweitat. This tribe receives no mention either by Burckhardt (1822) or by Robinson (1841). Thus, far we have intentionally refrained from integrating the history of the Gebeliya Tribe with the common history of all the other tribes, since the Gebeliya tribe, despite being one of the oldest and perhaps even the oldest in the region, is exceptional insofar as its history is concerned. On the beginnings of the Gebeliya, Burckhardt (1822) remarks: "To the true Bedouin tribes above enumerated, are to be added the advena called Djebalye, or the mountaineers. I have stated that when Justinian built the Convent [St Catherine], he sent a party of slaves, originally from the shores of the Black Sea, as menial servants to the priests." Also, according to Robinson (1841, p. 200) and Nandris (1981, p. 56), the evidence for the existence and complexity of the Gebeliya tribe already appears in the writings of Eutychius, Patriarch of Alexandria, in the ninth century. This Patriarch relates how the Pope's emissary was dispatched along with 100 males and their families from among the Byzantine servants, and another 100 families from Egypt, all of them sent to the Santa Katharina Monastery to guard it. The Gebeliya apparently assumed the Moslem faith at the time of the Caliph Abd-el-Malik (625–705), even though a few members continued to retain the Christian religion until the middle of the eighteenth century (Burckhardt, 1822, p. 564). Nandris (1981) maintains that the population from which the first Gebeliya originated (the Vlah, as recorded in the writings of travelers and explorers) still exists under that name in southern Europe. It is clear, then, that the first members of the Gebeliya tribe reached South Sinai in the sixth century and as a tribe has survived until the present. This remarkable survival as a cultural entity is attributable to two main factors, namely: (a) intensive economic support by the monastery authorities, which enabled them to withstand severe drought periods; and (b) cultural isolation from the rest of the South Sinai tribes.

### Genetic evidence of the origin and relationships between South Sinai Bedouin tribes

According to the historical evidence, we believe that the tribes of South Sinai may be divided into four main biological units: the first includes the Sawalcha, Gararsha and Awlad Said tribes; the second, the Muzeina tribe, the third, the Aleigat and Hamada tribes, and the fourth, the Gebeliya tribe. The Haweitat, Beni-Wassal and Ahali Et-Tur are probably independent biological units. The regional origin of all these tribes, with the exception of the Gebeliya, according to most investigators, is the Arabian Peninsula. Yet, this view seems to us problematic, in that some of the South Sinai tribes did not arrive directly into Sinai from Saudi Arabia, nor did they arrive as a single homogeneous group. An example is the settlement of the Awlad Said tribe in the Sinai as reported by Ben-David (1978). According to this author, the Abu-Zohar clan, for example, was the first to reach South Sinai from the Arabian Peninsula, whereas the Abu-Alaj clan wandered in the North African dunes before reaching South Sinai from Tunisia. Assuming this information is essentially reliable, it suggests the possibility that the various tribes had incorporated groups not originally present in the 'mother' population'. Hardly a tribe in South Sinai had not at one time or another assimilated extraneous families (see, for instance, Figure 4.2 depicting the structure of the Muzeina tribe).

There can be no doubt that slaves were brought to Sinai from the Sudan, about which we have written documentation.

Thus "But poor as they are, some of them, especially the Gararsheh, possess Negro slaves who look after the camels" (Palmers 1871, p. 84). And "Between three hundred and four hundred Negroes live in Sinai, the majority near Tor. Sheikh Eid commented that forty or more years ago, each big Sheikh had Negroes as bodyguards and slaves" (Field, 1952, p.129).

Such admixtures pose the question whether the individual tribes may be defined as homogeneous biological units. We attempt to answer this question subsequently. First, however, we shall examine the genetic evidence for the origin of, and the interrelations between, the various tribes.

The first studies to examine the frequency of genetic markers in some blood group systems, haptoglobins and transferrins, in Bedouin populations of South Sinai were carried out by Kaufman-Zivelin (1971) and Bonne et al. (1971). The data included the frequency of haptoglobin phenotypes in the Bedouin population of South Sinai. Kaufman-Zivelin (1971) concludes that the Gebeliya tribe is distinct from the rest of the mentioned tribes because in it the frequency of the Hp1-1 phenotype is

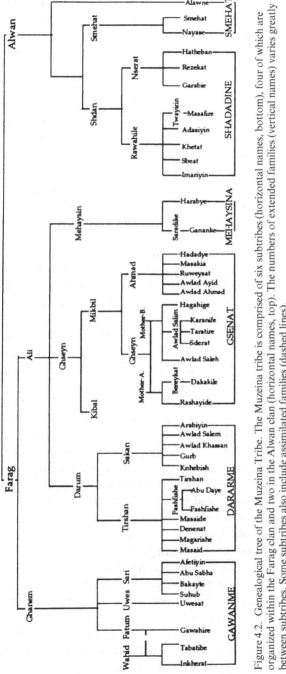

Figure 4.2. Genealogical tree of the Muzeina Tribe. The Muzeina tribe is comprised of six subtribes (horizontal names, bottom), four of which are organized within the Farag clan and two in the Alwan clan (horizontal names, top). The numbers of extended families (vertical names) varies greatly between subtribes. Some subtribes also include assimilated families (dashed lines).

probably close to zero, while the frequency of the *Hp2-2* attains 63% and is greater than in the other tribes. As for the *Hp2-1* heterozygote, the differences between the tribes are less marked. Calculation of the *Hp1* gene frequency for the various tribes shows it to be lowest in the Gebeliya (0.19) and highest in the Aleigat (0.51). Kaufman-Zivelin (1971) also computed the observed frequency of the haptoglobin types versus the expected frequency according to the Hardy–Weinberg law for a panmictic population of unlimited size, and without selection, migration or mutations. She found that discrepancy between the observed and expected frequency occurred only in the Gebeliya tribe. She likewise compared the frequencies of the three phenotypes between tribe pairs, and clearly showed that the Gebeliya tribe differs significantly from all the other tribes and also that the Muzeina tribe differs significantly from the Aleigat. Comparison of ABO blood groups in the various tribes has shown that the frequency of the *q* gene *is* high in the Gebeliya tribe relative to that in the other tribes, whereas in this same tribe the frequency of the *p* gene is comparatively low. The "exclusivity" of the Gebeliya from the rest of the tribes was verified. The frequency of the *p* gene in the Muzeina tribe (0.572) was significantly higher compared with that in the Gebeliya (0.309) or the Aleigat (0.366) tribes. Kaufman-Zivelin, in her Bedouin study (1971), attempted to ascertain whether there was any correlation between the frequency of the *HP1* gene in the various tribes and its frequency in the geographical regions from which the tribes purportedly originated. Her data show similarity to that in Arab and other Bedouin populations in the Near East. The Gebeliya tribe in South Sinai manifests a frequency of the *HP1* gene (0.18) that is much lower than usual in Europe (0.34–0.37) or, excepting Egypt (0.21), in Africa (0.40–0.63). Kaufman-Zivelin also addressed the question as to whether the Gebeliya tribe represents an amalgamation of African and European populations. She chose a large number of genetic markers whose frequency differs in African and European populations, and concluded that both African and European influences are discernible in their (Gebeliya) genetic systems and that deviation from the expected gene frequency, as presented by the two "mother populations", is putatively the result of selection processes or genetic drift. She also concluded that the Gebeliya is the only tribe that stands apart from all the other tribes in *Hp1* and that a comparison of its frequency among the Bedouins in South Sinai and neighboring populations strengthens the assumptions regarding the geographical origins of the various tribes, as previously indicated.

Contemporaneously with the study of Kaufman-Zivelin (1971), an extensive study by Bonne and coworkers (1971) appeared on heritable blood factors in the Bedouin. In all, 297 individuals (280 males and 17 females) were examined (excluding family ties of the first order). The sample was divided into four main groups, namely, Gebeliya (95), Aleigat (50), Muzeina (53) and the remaining Towara tribes (99). The authors conclude that the Gebeliya tribe differs significantly from the rest in numerous genetic systems and that, with few exceptions, there are no significant differences between the other tribes. This is the reason why the data are given only for the Towara and the Gebeliya. Bonne *et al.* (1971) also note that:

> In most systems the Towara Tribes agree with what is known of other neighboring peoples, and for the blood groups in the strict sense, they greatly resemble tile Arabs of the Arabian peninsula. The main differences are the high frequencies of the Rh complex *cde* (r) and the genes *K* and *PGM1-1*. The Jebeliya on the other hand, differ very markedly and significantly not only from the other Sinai Bedouins but, from all other neighboring populations ... In general, the prominent negroid features observed in the Jebeliya call for a reappraisal of their historical and ethnic background (p. 407).

Ben-David and his associates (1983) examined the frequency of the *T* autosomal dominant gene among a number of Bedouin tribes. The ability to taste the compound phenylthiocarbamide (PTC) is generally attributed to a dominant autosomal gene, called *T*, whose recessive allele is *t*. Homozygote dominants (TT) and heterozygous (Tt) individuals are able to taste PTC, whereas the recessive homozygote individuals (tt) are unable to do so. The results show that the Gebeliya tribe differs significantly from all the other tribes with which it was compared with one exception, the Awlad Said. The Muzeina tribe also differed from all the other tribes, without any exception. Furthermore, the high frequency of the *T* gene encountered in the Gebeliya tribe is typical also for many African societies (see Mourant *et al.*, 1976).

The biological data studied show that: (a) the Aleigat and Hamada tribes are related; in genetic systems such as sensitivity to PTC, P blood group system and such, they show almost complete identity and can therefore be treated as one group; (b) the Muzeina tribe may be regarded as an independent biological unit because it does not show genetic proximity (albeit judged by blood markers only) even to the Aleigat tribe with which it purportedly had maintained marital relationships; (c) the Gebeliya tribe also appears to be an independent biological unit which, like the Muzeina tribe, shows little or no genetic relatedness to any of the other tribes in South Sinai. Unlike the Muzeina, however, it displays extreme variability in some of the genetic systems. This variability, or, "instability", stems probably from the fact that previous field investigators failed to discriminate between the Awlad Gindi subtribe whose origin is probably Egyptian, and the other subtribes which originated from Bedouin tribes of the Arabian peninsula, perhaps even from surviving slaves brought by Justinian to the Santa Katharina monastery from Europe in the sixth century.

The frequencies of the genotypes of the blood groups do not show any deviation from the Hardy–Weinberg law, with one exception in spite of the fact that we are dealing here with small populations with non-random matings. The one exception is the Gebeliya tribe, although the reason for this is not clear. Why a "genetic balance", expressed by blood group systems, appears to be maintained within the South Sinai Bedouin groups is still unclear.

### Social structure

There are three large "social units" among Bedouins: (a) "tribal suprastructure", (b) "tribal affiliation", and (c) "tribe". In the majority of Bedouin groups in northern Sinai and in Israel, the biological ties between and within the tribes find expression in the tribal suprastructure organization. The tribes of South Sinai, however, diverge somewhat in this respect. While they all belong to one super-tribe organization (Safef in Arabic), known as the Towara ("Tor" in Arabic meaning a mountain, thus designating the topographic nature of South Sinai), the ties between the tribes are based mainly on common and defined geographical localization and not on any "blood" relationship such as is customary at the tribal suprastructure level in most Bedouin groups (Marx, 1974; Ben-David, 1978). Baily (1977, p. 246) notes that "the tribal suprastructure is the nationality of the Bedouins."

The inception of most of the major Bedouin tribal suprastructures in Sinai and the Israeli Negev started with a limited number of tribes, which initially consolidated in the Arabian Peninsula and then migrated north and northwest into settled land. In the course of this migration moving from one region to another the group of tribes assimilated local families, factions and at times even entire tribes, to form an expanded tribal suprastructure. The founding of the majority of large tribal suprastructures of Bedouins in Sinai and the Negev was in the Arabian Peninsula (for example: Tarabin, Azazme, Tayaha). The Towara suprastructure, which includes most of the Bedouin tribes of South Sinai, unlike the other large tribal suprastructures was formed by a conglomerate of discrete tribes with no genealogical or historical ties between them. The Towara in South Sinai had been organized to form a united body, powerful enough to counterbalance the other tribal suprastructures, especially the Tayaha which sought to encroach on its living spaces. However, apart from the latter common purpose, the Towara exerted no influence on occurrences between and within tribes in the South Sinai.

In South Sinai there also exists a "framework" of tribal affiliation or alliances, which is all organization between tribes within the tribal suprastructure. A similar organization exists also among the Negev Bedouins. Baily (1977) called the latter organization "EL-Bateh" or the "EL-Fahed". In the South Sinai affiliations the ties between tribes are primarily on a sociopolitical level; in the Israeli Negev affiliation, however, the ties are essentially blood relationships, through a common ancestor. According to Baily (1977), the role of the latter tribal affiliation is to form a framework for social intercourse, such as intertribal marriages, or cooperation in pilgrimages to the graves commonly considered holy, usually tombs of sheikhs known for their "supernatural" powers, and also for economic cooperation (particularly in smuggling activities). The South Sinai tribal affiliations are called after the leading tribe in the group: the Sawalcha affiliation includes also the Awlad Said and Gararsha tribes, and the Aleigat affiliation includes also the Muzeina and Hamada tribes. The Gebeliya, Haweitat and Beni-Wassal tribes are not included within either coalition. The present tribal affiliations are not completely parallel with the biological units determined earlier on the basis of the different historical background of each tribe.

The third-ranking social unit from the standpoint of size is the tribe, or El-Ashira. According to Baily (1977) a tribe forms in one of two ways: (1) a subtribe increases in size and demands independence from the mother tribe in order to advance its special interests; or (2) a certain group assimilates other groups and together these come to comprise a tribe. Marx (1974, p. 58) has this to say about the Israeli Negev Bedouin tribe:

> The most extensive political group at present is the tribe (ashira: plu. ashair) and the sub-tribe (ruba; plu. rub'oa). The tribe was formerly the group that made battle. Today it is the administrative unit. At the head of such a group stands an elected leader, called the Sheikh ... the Sheikh is now the primary mediator, between the sub-tribes and the authorities – a fact which lends him considerable power ... the tribe also had an appreciable degree of continuity, and this is apparent from the fact that this is the smallest group in the customary genealogies which describe the relationships, between all the tribes of the tribal suprastructure. The tribe is not a group possessed of land properties and its boundaries are not so clearly demarcated as those of branches but generally one can say that the groups which comprise the tribe own adjacent lands which are utilized as areas of grazing.

## Reconstructing marriage pattern among South Sinai Bedouins

### *Marriages outside the tribe*

Despite the numerous testimonies as to the entry of foreign families into the South Sinai Bedouin tribes, there is little concrete evidence to this effect at present and even in the two preceding generations. The fact that the region of South Sinai is sealed by topographic barriers apparently has precluded regular traffic or communication with neighboring peoples. Thus, only 0.4–2.0% of the marriages in the tribe were with women from outside the tribe. Immigration from outside the Towara into the South Sinai Bedouin tribes is extremely small (usually less than 0.5%).

Some tribes do display a stronger tendency to take a wife from outside their Towara tribal suprastructure (e.g., the Gararsha) than others (e.g., the Muzeina). Even within the tribe there are parts (e.g., the subtribes), which display different tendencies in this regard. The geographical location of a subtribe apparently plays an important role in this respect. As mentioned, the Gsenat and the Dararme subtribes of the Muzeina tribe display such disparity. Thus, the former which lives in the center of the tribal territory, and has neighboring subtribes to the east, north and south, hardly accepts brides from outside the Towara, whereas the Dararme subtribe, which occupies the periphery of the tribal territory shows greater tendency for such marriages. Nir (1987) presents somewhat different data in these matters although he also emphasizes the significant trend for intertribal marriages. According to Nir (1987), the Gebeliya is the tribe with the highest frequency of first cousin marriages and the least with outside tribe marriages ( <1%). In contrast the Awlad Said tribe accepts women from practically all the tribes of the Towara and at rates greatly exceeding those of the Gebeliya and Muzeina tribes combined. In the Muzeina tribe, according to our data, the rate of female inflow averages about 3% for each of the last three generations. Nevertheless, there are differences also in the subtribe level.

### *Marriages within the tribe*

Undoubtedly genealogical ties between the subtribes exert a direct influence on the rate of female exchange between them, as illustrated in Figure 4.3. First, we note that a large proportion of the females taken as wives come from within the subtribe. For the Muzeina tribe the relevant percentage is approximately 77% (average for all subtribes). Of the remaining 23% of the marriages, only about 2.4% derive from other tribes or from

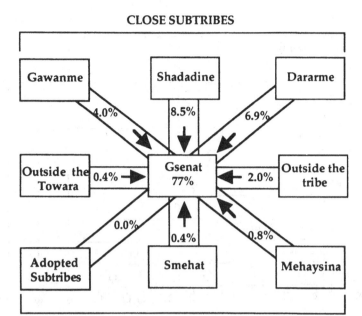

Figure 4.3. Women exchange in marriage between the Gsenat subtribe and other subtribes of the Muzeina Tribe in South Sinai. The rate of "foreign" women entering the subtribe Gsenat is 23%; 77% are marriages with women from the same subtribe.

outside the tribal suprastructure, the Towara, the remainder (20.5%) coming from neighboring subtribes. In the Gsenat subtribe, for instance (Figure 4.3), 19.3% derive from related subtribes and only 12% from remote subtribes (Smehat and Mehaysina). These data suggest that, at, least, in some cases the subtribe is the social unit, or alternately, a group of subtribes deriving from a common ancestor is such as, for example, the Gsenat, Dararme, Shadadine and the Gawanme subtribes, all tracing back to an ancestor called "Farag" (Figure 4.2). Our next step is to try to reconstruct the genealogical depth characterizing each type of marriage.

### Hams (blood feud group) marriages

As noted above, these marriages are limited to between cousins. Figure 4.4 describes several such possibilities, e.g., between individuals 1 and 2 who are first cousins, between individuals 1 and 3 who are second cousins, and between individuals 1 and 4 who are third cousins. Marriages between

78    E. Kobyliansky and I. Hershkovitz

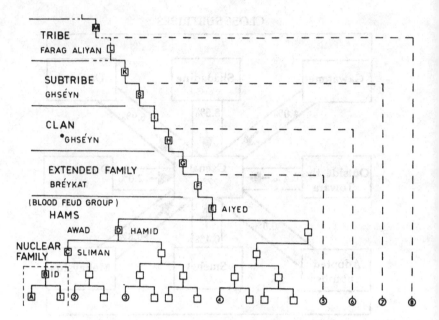

Figure 4.4. The genealogical depth of the various marriage types in the Hamid
Aiyed family. A–M, ancestors and their place within the social framework of the
Muzeina tribe; 1, Ego; 2–8, Potential wives for marriages and their genealogical
relation to Ego. • One of Ghese'yn's grandsons whose descendants form part of
the Gsenat subtribe (see Fig. 4.2.)

cousins are in most cases marriages with offspring of the father's brothers
although there is evidence of other combinations as noted previously. The
genealogical depth ranges between two generations for marriages between
first cousins to four generations for marriages between third cousins
(Figure 4.4).

### Marriages in the extended family

Here we are dealing with marriages between two individuals of the same
family, e.g., between individuals 1 and 5 in Figure 4.4. The common
genealogical depth for the married couple reaches down to seven generations (Figure 4.5). The remotest ancestor is Breykat (Figures 4.2 and 4.4).
Clan marriages are between individuals belonging to two different families
related by a close genealogical tie, for example, a marriage between individuals belonging to families sharing a common ancestor. Thus, for the
common ancestor Ghse'yn we have the following families: Rashayide,

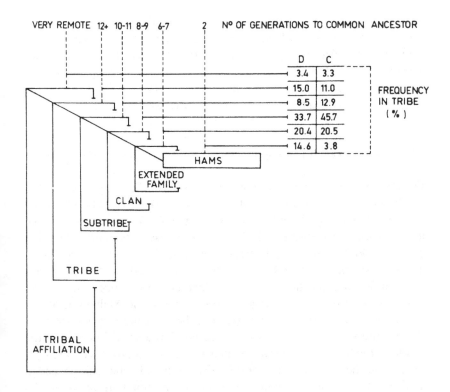

Figure 4.5. Number of generations of the mates from a common ancestor in various types of marriages.

Breykat, Awlad Saleh, Awlad Salim and Hagahige (Figure 4.2). Another example is marriage between individuals belonging to families originating from the common ancestor Sari; such families are the Suhub, Bakayte, Abu Sabha and Afe'tiyin (Figure 4.2). As observed in Figure 4.4, we are dealing with a marriage between individuals 1 and 6. The common genealogical depth for this couple is nine generations.

### Marriages in the subtribe

We here consider marriages between individuals of families who belong to the same tribe, and are genealogically linked through one of the two sons of the tribal founder, either Farag or Alwan (Figure 4.2). Such, for example, is the marriage between individual 1 and 8 in Figure 4.4. The genealogical depth common to these two marrying individuals ranges between 10 and 12 generations (Figures 4.4 and 4.5). In sum, as noted in the preceding, a precise schema which describes the structuring and development of the tribe from its inception to the present enables one to trace the common "biological" background of each marrying couple. The only information needed to ascertain the genealogical depth of parents is the name of the extended family from which a wife derives.

### Demographic considerations

The large tribal suprastructures of the Negev are the Tarabin (25 tribes numbering 33 062 individuals) and the Tayaha (29 tribes numbering 27 063 individuals). Compared with these groups, the South Sinai Towara tribal suprastructure is rather small, the maximum estimation in 1976 being about 10 000 individuals in 10 tribes. The growth rate of a tribal suprastructure per year between 1929 and 1968 was about 0.9%. In the Towara tribes in South Sinai the number of individuals ranges between 151 and 1099 per subtribe. In the Muzeina, for instance, the Mehaysina is the smallest subtribe, with 261 members, while the Dararme is the largest, with 785 members. The number of extended families also differs among the tribes. In the Muzeina tribe we counted 45 extended families (Figure 4.2), whereas in the smaller Gebeliya tribe there were 24. There are extended families, which number scores of members, at times upwards of 100 individuals, whereas very small extended families may be comprised of few individuals, sometimes even fewer than 10. In the South Sinai tribes the number of nuclear families per tribe ranges from 88 in the smallest tribe, namely the Haweitat, to 880 in the largest tribe, the Muzeina; the number of families per subtribe ranges from 71 in the smallest subtribe of the Muzeina to 267 in the largest subtribe of this tribe.

Despite the major censuses taken in the Sinai in 1968 and in 1975–6 by the Israeli authorities, there are no exact data on the age distribution within the Bedouin tribes of South Sinai. The age pyramids for Bedouin groups from the Sinai, according to Ben-David (1978) and from the Negev according to Muhsam (1966), show that a major part of Bedouin in the South Sinai (44%) and in the Negev (49%) is concentrated in the low age group of 0–14 years (Table 4.1).

The same trend is true also in Arab villages in Israel (Muhsam, 1966). In the South Sinai and Negev Bedouins, after age group 0–14 the relative proportion of each age group diminishes drastically (Table 4.1), a trend in contrast to that of corresponding age groups in other populations. The males of 60+ years in the South Sinai and Negev Bedouin populations

Table 4.1. *Percentage distribution of the Bedouin in South Sinai[a]
and the Israeli Negev[b] by sex and age*

| | Males | | Females | | Both sexes | |
|---|---|---|---|---|---|---|
| Age group | Sinai | Negev | Sinai | Negev | Sinai | Negev |
| 0–14 | 23.1 | 25.0 | 20.8 | 24.5 | 43.9 | 49.5 |
| 15–29 | 10.7 | 11.1 | 9.2 | 11.1 | 19.9 | 22.2 |
| 30–44 | 9.8 | 8.2 | 9.9 | 7.8 | 19.7 | 16.0 |
| 45–59 | 7.9 | 4.2 | 4.6 | 4.0 | 12.5 | 8.2 |
| 60+ | 2.7 | 2.0 | 1.3 | 2.1 | 4.0 | 4.1 |
| Total | 54.2 | 50.5 | 45.8 | 49.5 | 100.0 | 100.0 |

[a] Ben-David, 1978; [b] Muhsam, 1966.

comprise only 2.0–2.7% while in the Israeli male population they are 12.3%. Relatively many fewer South Sinai Bedouins live to old age, probably due to their harsh environment.

As for fertility rate, we calculated a crude birth rate (CBR) of 38.7 live progeny per 1000 individuals of population. Ben-David (1978), who made an analogous computation for all the Bedouin tribes of South Sinai, based on the data of the 1968 census and on a mortality rate approaching 45% (of all children from 0 to 14 years), obtained also a CBR of 40 live births per year per 1000 individuals. The GFR value is 228.9, estimated as annual birth rate (122) divided by number of fertile women (533) multiplied by 1000. The estimated value for the total fertility rate (TFR) among South Sinai Bedouins was 6980. We find that the fertility rates (DBR, GFR, TFR) in the Bedouin society of South Sinai are relatively high compared with those in populations of industrialized countries, but are comparable to those in neighboring Arab countries, or third world countries. The fertility in the Bedouin society, on the basis of the data collected in our survey (Kobyliansky and Hershkovitz, 1997), shows that by the time of menopause (age 45+ years) the Sinai Bedouin woman has had an average of 5.50 living children, the Negev Bedouin woman, 5.80 children, and the Arab village woman, 6.04 children.

The average fertility period of the South Sinai Bedouin female is about 18 years (range 16–44 years), with overall fecundity of almost seven children. The average birth interval of a married Bedouin woman in South Sinai is 2.4±0.4 years. In the Bedouin populations of the Negev, the duration of the fertile period in women is 18.6 years, live births occur every 2.5 years, and the overall fecundity is 7.5 children per woman (Muhsam, 1966). We also found that among young Bedouin fathers (21–35 years), the mean marrying age had been about 20 years, and in older age groups

82    E. Kobyliansky and I. Hershkovitz

(36–45 years), it was about 26 years, calculated according to the age of the first child. These data suggest a change in the marrying age of males after the year 1967, during the Israeli presence in the territory. Apparently a more rapid accumulation of property occurred after 1967, which enabled earlier marriage by South Sinai Bedouins, and consequently a greater mean number of children per family.

Regrettably, we were unable to compute any mortality indices (crude death rate and infant mortality rate) that could be comparable to those for other groups. From 55 families, where the mothers were relatively young (15–29 years) 13 reported at least one dead child, with the number of boys almost double the number of girls. In 42 families where the mothers were much older (35–44 years), 35 families already reported on at least one dead child in the family, the ratio between male and female remaining the same.

The mortality curve for the Bedouins in South Sinai, like that for many other populations, is U-shaped. Thus, among the Bedouins the mortality rates in the low age groups (0–14 years) are exceptionally high, but thereafter (15+ years) the rates decline rapidly, remaining relatively low till the end of the third decade of life when they rise sharply again albeit not to the same extent as in the early ages.

The state of health among South Sinai Bedouins has until recently received relatively little attention (Levi, 1978). Nonetheless, the available data suffice to emphasize how important this factor was in regulation of Bedouin population size in South Sinai before 1967. Major plagues, about which we have the testimony of travelers as well as an abundance of native folklore (Levi, 1987), accounted for a very slow total increase of population over time, even diminishing it markedly at intervals. Infants who became ill or mothers who contracted a postpartum infection, were in most cases doomed to death. The harsh living conditions – cold winters in a wind-exposed tent, and hot summers with little shelter, coupled with a meager and inadequate diet, rendered the population, particularly the young, prone to numerous diseases. Undoubtedly, poor nutrition and harsh living conditions also induced lengthening of the lactation period in women, and hence the overall fertile span was shortened (Frisch, 1978).

The replies to a questionnaire among the South Sinai Bedouins indicated that the mean marrying age was about 23 years for a male and 17.5 years for a female. Nir (1987) found that the marrying age for females ranged from 16 to 18 years, and for males between 23 and 25 years in this same population. According to Muhsam (1966), among the Negev Bedouins, some 50% of the men were married by the age of 20 and 75% were married by age 24 (mean 22.4 years); among Negev Bedouin women, 50%

were already married by age 17 and 80% were married by age 19 (mean 18.4 years).

The age difference between South Sinai Bedouin males and females at the time of marriage was between 2 and 6 years; for Negev Bedouins it was about 4 years (Muhsam, 1966). Yet there are some males who marry at a relatively advanced age, and sometimes they do so with young women despite the general social disfavor of marriages between an older man and a young woman. Of interest is that relatively few males (less than 4%) marry women older than themselves.

The early marrying age of females in the Bedouin society theoretically enables them to produce more progeny during their fertile period. There is some question as to whether a correlation exists between marrying age (except in later years), either of males or females, and fertility (Revelle, 1968). Most studies, however, suggest that there is no such correlation (Benedict, 1972).

In regard to polygamy, by Moslem law each Bedouin male can marry up to four women. Yet, a Bedouin man with three or four wives is rare (Muhsam, 1966). On the basis of genealogical records, we found varying rates of polygamy throughout the studied generations, with a general trend towards diminution of the polygamy rate in time. We have also found, as expected, that frequency of polygamy increased with the husband's age. The mean rate of polygamy obtained for the total population, based on questionnaire data, was 15%, which is close to that computed from genealogical trees, i.e., 12.1%. Ben-David (1978) reported 17% polygamous families in the South Sinai Bedouins. Nir (1987), on the basis of 1900 families, reported a 12.25% polygamy rate: 10.15% husbands with two wives, 1.68% with three wives and 0.42% with four wives, and maintained that only sheiks or wealthy and well-known men possessed more than two wives. Nir also gives the rate of polygamy in the various tribes: Aleigat 21.3%; Haweitat 12.9%; Gebeliya 7.5%; Sawalcha 21.1%; Awlad Said 11.1%; Hamada 5.5%; Gararsha 15%; and Muzeina 8.6%. The general trend from the data is that the larger the tribe the less the frequency of polygamous families. i.e., 8.6% within the Muzeina (3056 members) versus 21.1%, in the Sawalcha (401 members). Levi (1979) notes that of 81 examined polygamous families, 67 were bigamous, 10 trigamous and 4 quadrigamous. Taking all the Negev Bedouin tribes together, the mean percentage of polygamous (actually bigamous) husbands in the late 1940s was 7.7, with a range of 4.4–15.5% (Muhsam, 1966). Marx (1974) has maintained that polygamy is relatively rare among Bedouins, citing 9.2% for the Abou-Juoad tribe of the Negev, emphasizing that among the

Sheikhs, polygamy was not uncommon, especially when compared with the overall Bedouin population in the Negev.

In the South Sinai the mean age at which the Bedouin males acquire a second wife is 31 years. Muhsam (1966) estimates this age for Negev tribes as 35.8 years. He further calculates the mean age difference between a husband and his second wife to be 14.5 years, and the age difference between the first and second wife, about 10 years. The demographic data in this respect for both Negev and South Sinai Bedouins, are quite similar, so that the Bedouin male is likely to be some 11–14 years older at the time of his second than at his first marriage and usually his second wife is a woman not much older than was his first wife at her marriage.

Muhsam (1966) found that in polygamous marriages, only 80% of the husbands were older than their first wife, whereas in monogamous marriages the corresponding value was over 90%. He attributes these findings to the fact that in many cases Bedouin males were required to marry the widow of their deceased older brothers. Such "forced" marriages could lead to a situation where the first marriage of a younger brother might well be with an older woman, and consequently "trigger" the young brother to acquire a second youthful wife.

Levi (1979) maintains that the primary advantage of polygamy is that it enables the male to produce many offspring thus enhancing the "physical" strength of the family. Muhsam (1966), on the other hand, claimed that polygamy lends an economic benefit to the Bedouin male in that it enables him, in times of drought and aridity, to split up his herds and send them to different watering holes, each under the responsibility of one of his wives.

We found a negative correlation between husband and wife kinship and polygamy, namely, the closer the kinship between husband and wife, such as being first cousins (marriages within the Hams), the less the likelihood that the husband would acquire an additional wife. We cannot yet offer an explanation for this fact; the implications of this observation will be considered subsequently. An important, albeit controversial, issue is whether polygamy exerts an effect on the fertility of an individual woman. Theoretically, this should have no effect, for according to Bedouin tradition, a husband with more than one wife must share his nights equally among all his wives, each night with only one wife, without favoring one wife over another.

Second marriages, however, frequently result from inability of the first wife to produce many, or any, offspring. Benedict (1972) cited such "reasons" for the inverse correlation between polygamy and fertility. Most investigators concerned with the question, concur that polygamy is negatively correlated with fertility (see Benedict, 1972). It is of interest that

Muhsam (1966) noted that in Negev Bedouin polygamous families, sterility occurred more among first wives than in the monogamous families. However, his data also indicated that the percentage of infertile second wives was greater than that among first wives in polygamous families, something we did not encounter among the South Sinai Bedouins.

A further significant finding by us was that among the South Sinai Bedouins polygamous marriages were less fecund than monogamous marriages. Thus, we found, on the basis of Hams records of 12 families that the average number of living children per wife in a polygamous family was 3.5 (the same for first and second wife). According to our questionnaire data (based on 15 polygamous families), the average number per wife was 4.0 (for both the first and second wife). In the monogamous family, however, the average number of children per wife was 5.5. Hence the disparity in fertility between monogamous and polygamous families, relating to live progeny is on the order of about a 36% differential. According to the data of Muhsam (1966, p. 93), the difference was even greater in this regard, 32%, assuming no difference in child mortality within the two groups and that the age distributions of the wives in the two groups are comparable.

The act of "divorcing" is prevalent in Bedouin society, yet is not reflected in the demographic data (Ben-David, 1978). Levi (1979) noted that in a sample of 70 Bedouin families from South Sinai, he encountered 17% of remarried women, and he suspected that this percentage must even be as high as 30% in other parts of the Sinai where economic conditions were less stable. According to Ben-David (1978), among South Sinai female Bedouins aged 15–39 years there were hardly any of a "divorced" status, whereas at age 40 years and over the average was 5.3%. Among the Negev Bedouins, according to Muhsam (1966), the percentages for the corresponding age groups were 0.46% and 0.72%. Two main reasons account for the "disappearance" of divorcees from statistical tables, based on official population censuses. First, a divorced woman, if still considered to be in her fertile period, soon finds another husband because of a great paucity of women in the South Sinai Bedouin society; each woman who becomes "available" is virtually immediately "abducted" by one of the males in the tribe. Second, divorced women return to their parents' house, if they do not remarry, and receive anew the status of a daughter. In such cases, the head of the family, when queried by census takers, declares the divorced daughter as merely his daughter without giving her past marital history. The divorce rate generally becomes higher, albeit very minimally so, starting in the older age groups (40+ years), when the women are past their fertility period. The reasons for divorce in Bedouin society are numerous and include, for example, infidelity, delinquency in housekeeping, infertility. Recompense to the woman divorced is negligible, yet is fixed and inherent within Bedouin law. Children, if there are any, will always be granted to the husband.

There may be a relationship between frequency of divorce and the desire of an adult male to have children. Thus, a man who does not have children with a first wife will divorce her, or take a second wife, and if the second wife does not conceive, he may also divorce her, and remarry a third or even a fourth time. Hence, if the fault for infertility does not reside in him, he will ultimately succeed in having offspring. A Bedouin man without children is considered 'as good as dead', for he is then cast to the fringes of society, his influence in the social framework of the tribe becomes minimal, and his physical and economic safety become dependent on others.

The rate of widowhood increases in the age groups. From 0 in the Sinai Bedouin (less than 0.1% among the Negev Bedouin), at the age group of 15–19 years, to 4.2% (7.3% among the Negev Bedouin) at the age group of 50–59. In both the South Sinai Bedouins and the Negev Bedouins, the number of widowers is relatively small compared with the number of widows (Muhsam, 1966; Ben-David, 1978). The latter attributes this to the fact that the widowed male does not long remain widowed, that he attempts to accumulate the financial means for obtaining a new wife, regardless of his age, and especially if he has children by the deceased wife. In contrast, a mature widow, that is, one past her fertile period who usually also lacks property, has less chance of remarrying.

Bedouin society demands that the fertile widow remarries and returns to the reproductive pool of the society. Thus most widows in the South Sinai Bedouin society do remarry. Factors that until recently prevented substantial growth of the population have been primarily malnutrition and poor health. A paucity of physicians and lack of medical care on the one hand, and poor economic conditions on the other, led to a very high mortality of infants at times almost 25% according to our data and 45% according to Ben-David (1978, p. 120). The ratio of males to females in a generation is a vital measure of the demographic composition in a population. The average sex ratio of human births is approximately 105 in favour of males (Cavalli-Sforza and Bodmer, 1971). Data on the sex ratio in the South Sinai Bedouin population in the 1960s and 1970s have been provided by Ben-David (1978). According to his data, unusual phenomena occur, with the sex ratio changing dramatically with the age. The mean sex ratio for all age groups combined was 118.3. According to Nir (1987), the sex ratio for the South Sinai Bedouins, all ages, was 109. Ben-David has noted that the sex ratio among Sinai Bedouins differed completely from that reported in the literature, especially for Western societies, where the number of males is larger at birth but equals the number of females at about age 5 years, and thereafter diminishes with age relative to the number of females.

The data collected by us pertaining to sex ratios are similar to those obtained by Ben-David (1978). From Hams (blood feud group) records we learn that in the parent generation (18–44 years), the sex ratio average was 128.1 and in the offspring generation (0–17 years) it was 106.9. The differences in sex ratio between the two generations may at least partially be attributed to an artificial disappearance of females from the Hams records in the fifteenth generation, i.e., those who did not reach adulthood and marry. According to Ben-David's (1978) data, the sex ratio in the parents' generation was 111, and in the children' generation 108. In our study we tried to overcome the tendency of the Bedouins to "eliminate" females from their reports, as already mentioned by Muhsam (1966), Ben-David (1978) and Nir (1987), by cross-checking the data with different resources. The data on sex ratio among Bedouins of the Negev prior to Israel statehood, according to Muhsam (1966), indicated a mean of 102 in favor of males for all age groups combined. The sex ratio in infancy (0–1 year) was 106.

### Methodological comments

By "immigration" we refer to the entry into the tribe in every generation of "foreign" women who in time produce offspring. The phenomenon of male "immigration" is very rare and thus not considered in the following calculations. Women who emigrate out of the tribe or who enter it without producing progeny are excluded from this category. Random breeding in small groups of $N$ individuals results in a loss of heterozygotes at a rate of approximately $1/2Ne$ per generation, if we ignore the factors of immigration and selection. On the assumption of random union of gametes, the less the heterozygosity the larger the value of $F$ per generation. All formulas used in the present study are based on Li (1968, p. 305), except as noted otherwise. The increment in $F$ is on the order of: $F_i = 1/2Ne + [(2Ne-1)/2Ne] \times F_{i-1}$. Immigration rate $(m)$, on the other hand, will prevent the group from attaining homozygosity and will shift the gene frequencies of the group towards the mean value for the total population. Such a process will lead to a decrease in the value of $F$. In this new equilibrium the value of $F_{im}$ will now decrease owing to the intrusion of immigrants: $F_{im} = (1-M)^2 \times F_i$, (Wright, 1951). In fact, the above formula describes the basic relationships between the $F$ and $M$ values without the effect of selection.

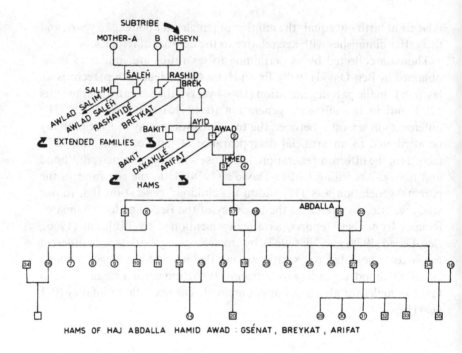

Figure 4.6. The origin of the Hams of Abdallah Khamid Awad and its place within the genealogical tree of the Muzeina Tribe.

Marital patterns constitute an important indicator as to how the genes are aggregated in the human genotype. Marital patterns are quite variegated in the various world populations. The reasons for such variability are apparently associated with biological, demographic, social, cultural and other factors.

In Bedouin society we recognize a number of marital "types" which, apart from their socioeconomic implications, also have a reasonably well-defined biological significance. It may be recalled that the marital "types" recognized thus far were based on the Hams (blood feud group), extended family, clan, subtribe, and tribal marriages. The main criterion used to define these marital "types" was the common biological background or depth of the two mates, i.e., the number of generations back to a common ancestor of the mates (Figure 4.5). The manner in which we translate these marital "types" in biological terms is illustrated in three "schemes", first, the scheme of Hams of Haj Abdulla (Figure 4.6), second, the private case of the Hamid Aiyed family; and third, the "scheme" of the Muzeina tribe given in Figure 4.2.

We may consider immigration in Bedouin society on two different social levels, namely: (a) women taken for marriage from outside Towara tribal suprastructure, and/or from outside her husband's tribe, but from one of the other Towara tribes; and (b) women taken for marriage from her husband's tribe.

The data presented therein pertain mainly to computations made for the current generation of the Muzeina tribe only; they do not necessarily reflect the situation prevailing in earlier generations.

### *Method of F computation*

Path analysis, first proposed by Wright (1921), enables one to compute a coefficient of inbreeding once we know the biological relationships between parents. This coefficient is the correlation between uniting gametes and is computed by assigning a value of 1 to gene "A" and a value of zero to its allele "a" (or, in fact, any arbitrary value). The correlation is usually designated by $F$ or $f$. According to Wright's model, if $n$ and $n'$ are the number of generations in the pathway tracing back to a common ancestor of both parents of individual Z, then:

$$F \sum_{i=1}^{n} (1/2)^{n' + n + 1}$$

In societies where there is a long chain of arranged marriages tracing back to a depth of many generations it is possible that in individual K the "A" gene is already in the homozygous state. Consequently, in every calculation of $F$, one needs to take into account also the $F$ values of the preceding generations. The fact that K is already a product of a long line of marriages within very small groups, increases the probability that other individuals will receive genes which are identical by origin from K by a factor of $F_A/2$. Therefore, all the marital loops linked to a common ancestor require multiplication by a factor of $1 + F_A$.

In our calculations we recognized a different $Fxe_i$ for each of the five conjugal types in the South Sinai Bedouins, thus: $Fxe_1$, for marriages within the Hams (blood feud group); $Fxe_2$, for marriages within the extended family; $Fxe_3$ for marriages within the clan; $Fxe_4$, for marriages within the subtribe; $Fxe_5$, for marriages within the tribe. On the basis of the structuring of marital patterns in the tribe, we computed for each $Fxe_i$ the genealogical distance, in generations, between the parents and their common ancestor.

The genealogical depth of the Muzeina tribe is estimated to be about 16 generations. We further assumed that the parental pairs on which there is no genealogical information, distribute proportionally into our five recognized marital categories; these comprise about 4.4% in the fifteenth generation according to the Hams blood feud group records. The computation of the suitable $Fxe_i$ was done according to the formula:

$$Fxe_i = PFxe_i + \sum_{i=1}^{n} [(1/2)^{n' + n + 1}]/100 - u$$

where $P$ is the proportion of the particular marital pattern within the total marital patterns in the sample, and $u$ is the percentage of cases per generation of parent pairs for which no adequate genealogical data were available to us. The final value of Fxe is obtained by combining all the computed $Fxe_i$ values:

$$F\bar{x}e = \sum_{i=1}^{5} Fxe_i$$

*Marriages within the Hams (blood feud group) ($Fxe_L$).* This category includes four main marital types. Marriages between first cousins are marriages to the children of the father's brother or those of the mother's sister. Of the total marriages within the blood feud group recorded from our genealogical trees (for the Muzeina only), 58.5% were marriages with offspring of the father's brother and only 18.8% with offspring of the mother's sister. This datum reflects the trend of cousins for marrying the children of the father's brother, but is not entirely free of error because the Bedouins are reticent about reporting marriages with the offspring of the mother's sister as marriages between cousins. For the same reason, there is in fact some question about the true marital rate of first cousins within the tribe. According to the studied genealogical trees, first-cousin marriages (of both types) probably do comprise about 77.3% of all marriages in the Hams. Second-cousin marriages account for about 15.1%, and third-cousin marriages some 2.0%. Marriages between relatives of two different generations are about 5.6%. It should be noted that the greater the genealogical depth, the greater the error in the estimates of the frequency rates of the marital types.

The contribution of different marital types within the Muzeina and their frequencies within the population to the $Fxe_i$ value are: First-cousin marriages: $(1/2)^5 \times 0.773 = 0.024156$; Second-cousin marriages: $(1/2)^7 \times 0.151 = 0.001179$; Third-cousin marriages: $(1/2)^9 \times 0.02 = 0.00003906$; cross-generational (uncle–niece) marriages between relatives: $(1/2)^6 \times 0.056 = 0.0008749$, with a combined value of 0.026248.

Since in the Muzeina tribe marriages within the Hams are only 14.6% of all marriages the relative contribution of this marital pattern to the $F$ value of the population is $Fxe_1 = 0.026264 \times 0.146 = 0.003834$.

*Marriages within the extended family (Fxe$_2$)*. The common genealogical depth for a pair of parents extends to 6–7 generations. We elected to use the latter datum albeit significantly reducing the obtained $F$ values. The frequency of this marital type accounts for 20.4% in the 15 (parent) generation. Thus, $Fxe_2 = (1/2)^{13} \times 0.204 = 0.0000249$.

*Marriages at the clan level (Fxe$_3$)*. The genealogical depth here is 8–9 generations and the frequency of this marital type is 33.7%. Hence, $Fxe_3 = (1/2)^{17} \times 0.337 = 0.0000026$.

*Marriage within the subtribe (Fxe$_4$)*. The genealogical depth amounts to 10–12 generations: frequency 8.5%. Thus: $Fxe_4 = (1/2)^{21} \times 0.085 = 4.053 \times 10^{-8}$.

*Marriages within the tribe (Fxe$_5$)*. The genealogical depth exceeds 12 + generations: mean frequency is 15%. Hence, $Fxe_5 = (1/2)^{25} \times 0.15 = 4.4703 \times 10^{-9}$.

The overall value of $Fxe_{1-5}$ obtained by us (0.0038615) comprises almost entirely (96.3%) the effect of first-cousin marriages. The contribution of all the other marital patterns to the value of Fxe is negligible (3.70%). All the Fxe values refer, of course, to one generation only, for when inbreeding processes are allowed to carry into a greater depth of generations the contribution of the other marital patterns to the overall Fxe value also increases. It may be noted that the obtained value is probably lower than the true value owing to the probability of inaccurate information provided on cousin marriages with offspring of the mother's sisters as previously observed.

The use made of the Hams (blood feud group) records to compute the Fxe for earlier generations poses two main problems: (a) the deeper one delves into early generations (e.g., first, second, fifteenth generation), the smaller the value of the expression $n' + n + 1$, the latter indicating the number of individuals participating in the inbreeding loop linking all the common ancestors of individual Z (Figure 4.7). Therefore, the composition of the tribe, as described in the present study indicates a fixed situation for a given unit of time. The further back we proceed in the number of generations, the more distorted becomes the structure of the tribe as we recognize it and the marital patterns may assume a genealogy entirely different from that shown in Figures 4.4 and 4.6. The frequencies of the different marital types are (*inter alia*) a function of group size. In a sample of 60 cases of new cousin marriages within the Muzeina tribe, we found that 53% occurred in blood feud groups numbering in excess of 60 individuals, 31.6% in blood feud groups of 40–60 individuals and only 15% in

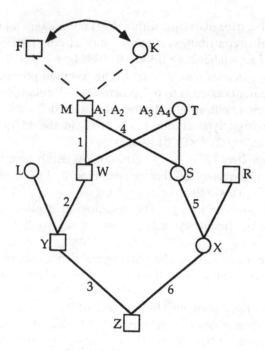

Figure 4.7. Genealogy of inbred individuals originating from one family (mates F and K).

blood feud groups of up to 40 individuals. Furthermore, in blood feud groups of less than 30 individuals cousin marriages were extremely rare. There is a clear correlation between Hams (blood feud groups) size and frequency of cousin marriages, but demographic reconstruction of the past history of the Hams did not enable us to evaluate fluctuations in population size or in the Hams size and therefore we must assume a fixed mean growth rate for all earlier generations. Consequently, we will consider the number of individuals in the Hams, in each generation, as stable, and the rates of marriages between cousins will also be estimated as stable since the number of cousins for each individual remains fixed in every generation. In order to compute optimally the overall coefficient of inbreeding for the South Sinai Bedouin society ($F$), after 15 generations, we took into account the following possible variables: Fxi, the inbreeding coefficient per generation; Fxe, contribution of the breeding patterns to the coefficient of inbreeding; and Fin, the influence of the genetic drift (depending on size of the $Ne$). Our final working formula, based on Li (1968) became:
$$F = [Fxi_{16}(1 + F_A)] \times (1 - M)^2.$$

There are three main parameters which contribute to a determination of the $F$ value. These are: the inbreeding coefficient for the present generation ($Fxi_{16}$), the contribution of earlier generations to the final inbreeding coefficient $(1 + F_A)$ and the effect of immigration during the last generation $(1 - M)^2$. The estimated $F$ value is 0.09802. It is noteworthy that both $Fxi_{16}$ and $F_A$ include (using various formulae in the different stages of calculation) the following variables: breeding patterns, genetic drift, differential fertility, sex ratio, and extent of polygamy.

### Immigration rates affecting the F value in South Sinai Bedouins

Since the tribe is the biosocial unit with which we are primarily concerned, it was deemed worthwhile to ascertain the effect which an average 3% migration late per generation would have on the $F$ value. The way of computation chosen for this purpose (after Li, 1968, p. 305) takes into account both the effect of inbreeding, as well as that of genetic drift. Thus, the $F$ value for the entire Muzeina tribe, when Ne = 846 and the migration rate is 3%, will be: $F = (1 - M)^2/[2Ne - 1)(1 - M)^2] = (1 - 0.03)^2/[1692 - (1692 - 1)(1 - 0.03)^2] = 0.009315$. The largest $F$ value obtained was for the tribal framework. The relationships between the migration rate on the one hand, and the effective size of the population on the other, are such that a rise in heterozygosity within the first two subunits, namely, the subtribe and the super-subtribe (stemming from an increased migration rate) is greater than the rise in homozygosity within the same groups, owing to reduction in their effective size.

The discrepancy between the value of $F$ for the sixteenth generation only ($F_{16} = 0.004297$) and the one obtained for all generations together ($F_{1-16} = 0.09802$) illustrates why $F$ values computed for different human societies are relatively small. Up to the early 1970s, the highest $F$ value computed for a human population ($F = 0.043$) was that for the Samaritans by Bonne (1963). Numerous studies (e.g., Salzano, 1972, on nine tribes of American Indians and Eskimo populations) along the same lines have led investigators to the conclusion that this value (0.043) was about the upper limit of $F$ possible in human populations. Subsequently, however, it was found that the upper limit can be somewhat higher ($F = 0.050$; Katayama et al., 1981). Among the many studies of the topic, the one by Spielman et al. (1975) is especially relevant to our present study, for the values, computed by those investigators for American Indian populations took into account past biological processes. Spielman and his associates noted that in such a computation the value of $F$ can turn out to be much higher than one might suppose. According to these authors (p. 367):

> ... it is difficult to escape the conclusion that except at loci with very high heterozygosity maintained by selection, identity by descent from ancestral alleles in Amerindian founders for a pair of randomly chosen alleles in a contemporary Amerindian tribe should be no less than 0.3 and may well be greater than 0.5.

## Conclusions

Most populations in the world today reside in more or less man-made environments. Relatively few have remained in the natural ambiance in which they had developed for hundreds or even thousands of years, continuing to maintain direct ties with it. One of the latter is the South Sinai Bedouin group which adopted the arid areas of the Middle East as its permanent residence, adapting to this extreme environment biologically, socially, and behaviorally. This chapter reviews possible links between cultural variables (historic, ethnographic, and demographic) and biological variables (organization of the genotypes into the various social frameworks).

Our study revealed that the present Bedouin tribes of South Sinai initially have immigrated from the Arabian peninsula to the Sinai starting more than 1000 years ago. In the ethnographic section we defined the organizational levels of the social institutions, namely, tribal suprastructure, tribal affiliation, tribe, subtribe, clan (khamula), extended family, and biological or nuclear family. Special emphasis was placed on the means of studying the Hams, the "blood feud" group. We have shown that the Hams, despite constituting only a judicial framework, serves as the best source for studying intratribal social dynamics and reconstructing marital patterns going back five generations. Our study revealed that the Sinai Bedouins comprise an endogamous society at the tribal level, 97% of marriages in the Muzeina tribe taking place within the tribe, with a certain preference for consanguineous marriages, 14.6% of the marriages being between first cousins.

In the demography section, we presented the main demographic characteristics of the South Sinai tribes. The average number of subtribes per tribe was found to be 4.2; the mean number of members per tribe was 1150 (range 183–3000); the number of nuclear families ranged from 56 in the smallest tribe to 880 in the largest. We presented and analyzed the demographic characteristics of the entire Bedouin population in the Sinai. The main findings were: (1) the population tended to be young, as many as 44%

being in the 0–14 years age group; (2) a high male to female sex ratio of 106.9; (3) as much as 6.9% of families beyond the reproductive period and without children; (4) a high proportion of remarried women (17%); (5) prevalence of second marriages; (6) a low marrying age both for females (17.5 years) and males (23.0 years); (7) a high mean number of living children per family (5.0 by the time of menopause); (8) overall familial fecundity of 6.98 children; (9) on average, wives of polygamous males bear fewer children (4.03) than do those of monogamous males (5.0); (10) male mortality higher than female mortality by 6% in the child population; (11) a clear link between consanguinity of parents (first cousin marriages) and mean number of children per family; (12) a low number of polygamous families (12.1%).

In the last part of this chapter we discussed the complex of factors which influence the genetic makeup of the Bedouin population. We relied here on the existing historical, ethnographic, and demographic information compiled in the previous parts. We considered the subject of "effective population size" and the factors which affect it directly, such as differential fertility, sex ratio, polygamy, temporary changes in population size and consanguineous marriages. In addition, we discussed the various ways of computing "effective size". Extensive study of the various mating types helped to illuminate certain aspects of "internal migration" within the tribes. We defined six mating types and computed their frequency in the various subtribes of the Muzeina. For each of these we computed the genealogical depth common to both mates. All this information pertaining to frequency of the various marital patterns on the one hand, and the common genealogical depth of the mates in each marital pattern on the other hand, enabled us to compute the relative contribution of the endogamous component (Fxe = 0.0038) to the coefficient of inbreeding, $F$. Thus we possessed most of the components for computing the general inbreeding coefficient $F$ for the most recent generation of the Bedouin population, using the series of equations developed by Wright in 1931.

The components at hand were: effective population size $(Ne) = 846$; migration into the tribe $(M) = 3\%$; marital patterns (Fxe) = 0.0038. We found that the $F$ value for the last generation equals $\text{Fxi}_{16} = 0.0043$. This value, however, did not take into account the contribution of preceding generations (FA) to the $F$. Ignoring a continuing process of inbreeding, taking place over hundreds of years within a small group, could crucially affect the value of $F$. Therefore, by a careful reconstruction of the demographic history of the Muzeina tribe, based on two different models of the dynamics of intertribal social processes, we evaluated the contribution of preceding generations (FA) to the $F$. The FA component, we discovered, produced a critical change in the magnitude of the inbreeding coefficient value $(F)$ and thus completely altered our thinking on the genetic structure of the Bedouin society. The new $F$ value obtained for the Muzeina tribe was 0.09802, a very high value that has rarely been recorded for any other human population.

## Acknowledgements

The authors would like to show their deep appreciation to all members of the research team who participated in the field study in South Sinai; Professor Baruch Arensburg, Dr. Meer Sarnet, Dr. Samuel Tau, Dr. Zion Cohen, Dr. Zvi Alter, Dr. Uri Car-Guy, Rachel Nefesh, Judith Blaushield, Eliezer Bergman and Shaul Gur-Lavi. For financial support of the research we are very thankful to the Schreiber Foundation, Tel Aviv University, and to the Dubek Company, Israel. We wish to thank the hundreds of Bedouins, children and their parents, who took part in the research and showed a lot of patience and understanding during the long period of the research. Much of the data collected were published in Kobyliansky and Hershkovitz (1997).

## References

Arensburg, B., Hershkovitz, I., Kobyliansky, E. and Micle, S. (1979). Southern Sinai Bedouin tribes: Preliminary communication on an anthropological survey. *Bulletin et Memoires de la Societé d'Anthropologie de Paris, t.6, serie XIII*, pp. 362–372.

Baily, Y. (1977). The Bedouins in Sinai. In: *The Desert*, ed. E. Zohar, pp. 240–247. Tel Aviv: Reshafim (in Hebrew).

Baily, Y. and Peled, R. (1974). *The Bedouin Tribes in Sinai*. Tel Aviv: Ministry of Defense.

Bar-Yosef, O. (1980). The Stone Age in Sinai. In *Kadmoniot Sinai*, ed. Z. Meshel and I. Finkelstein, pp. 11–40. Tel Aviv: Hakibutz Hameohad (in Hebrew).

Bar-Yosef, O. (1981). Neolithic sites in Sinai. In *Contributions to the Environmental History of Southwest Asia*, ed. W. Frey and H.P. Uerpmann, pp. 217–235. Beibette zum Tubinger Atlas des Vorderen Orientes, Reihe A, no.8.

Beit-Aryeh, Y. (1980). Early Bronze Settlement in South Sinai. In: *Kadmoniot Sinai*, ed. Z. Meshel and I.Finkelstein, pp. 295–312. Tel Aviv: Hakibutz Hameohad.

Ben-David, Y. (1978). *Bedouin Tribes in the South Sinai*. Jerusalem: Keshet (in Hebrew).

Ben-David, Y. (1981). *Gebeliya: A Bedouin Tribe in a Monastery Shade*. Jerusalem:

Kana (in Hebrew).

Ben-David, Y., Kobyliansky, E., Micle, S., Hershkovitz, I. and Arensburg, B. (1983). Taste sensitivity to phenylthiocarbamide in Bedouin of South Sinai. *Journal of the Israeli Medical Association* **3–4**, 57–58.

Benedict, B. (1972). Social regulation of fertility. In: *The Structure of Human Populations*, ed. G.A. Harrison and A.J. Boyce, pp. 73–89, Oxford: Clarendon Press.

Bonne, B. (1963). The Samaritans: a demographic study. *Human Biology* **35**, 61–89.

Bonne, B., Godber, M., Ashbel, S., Mourant, A.E. and Tills, D. (1971). South Sinai Bedouin. A preliminary report on their inherited blood factors. *American Journal of Physical Anthropology* **34**, 397–408.

Burckhardt, J.L. (1822). *Travels in Syria and the Holy Land.* London.

Cavalli-Sforza, L.L. and Bodmer, W.F. (1971). *The Genetics of Human Populations.* San Francisco: W.H. Freeman.

Field, H. (1952). *Contribution to the Anthropology of the Faiyum, Sinai, Sudan and Kenya.* Berkeley, CA: University of California Press.

Finkelstein, I. (1980). A Byzantian monestry at Jebel Sufsafeh-Southern Sinai. In *Sinai in Antiquity*, ed. Z. Meshel and I. Finkelstein, pp. 385–410, Tel Aviv: Hakibutz Hameohad.

Frisch, R.E. (1978). Population, food intake and fertility. *Science* **199**, 22–30.

Goren, A. (1980). The Nawamis in South Sinai. *Kadmoniot Sinai.* In *Sinai in Antiquity*, ed. Z. Meshel and I. Finkelstein, pp. 243–264, Tel Aviv: Hakibutz Hameohad.

Katayama, K., Kudo, T., Suzuki, T. and Matsumoto, H. (1981). Genetic studies in Tobishima: I. Population structure. *Journal of Anthropological Society of Nippon* **89**, 427–438.

Kaufman-Zivelin, A. (1971). The distribution and frequency of haptoglobins and transferrins of South Sinai Bedouin. M.Sc. thesis, Tel Aviv University.

Kobyliansky, E. and Hershkovitz, I. (1997). *Biology of Desert Populations South Sinai Bedouins: Growth and Development of Children in Human Isolates.* Liège: ERAUL 82, Etudes Recherches Archaeologique de l'Université de Liège.

Levi, S. (1978). *Medicine, Hygiene and Health Among the Bedouins of South Sinai.* Tel Aviv: The Society for the Protection of Nature (in Hebrew).

Levi, S. (1979). *The Bedouin Family of South Sinai.* Tel Aviv: The Society for the Protection of Nature (in Hebrew).

Levi, S. (1987). *The Bedouins in Sinai Desert.* Tel Aviv: Schocken (in Hebrew).

Li, C.C. (1968). *Population Genetics.* Chicago: University of Chicago Press.

Marx, E. (1974). *The Bedouin Society in the Negev.* Tel Aviv: Reshafim (in Hebrew).

Mourant, A.E., Kopec, A.C. and Domaniewska-Sobczak, K. (1976). *The Distribution of the Human Blood Groups and Other Polymorphisms.* Oxford: Oxford University Press.

Muhsam, H.V. (1966). *Bedouin of the Negev: Eight Demographic Studies.* Jeruselum: Academic Press.

Nandris, J. (1981). Tribal identity in Sinai. *Archaeology* **2978**, 56–57.

Nir, Y. (1987). The Bedouin tribes of Southern Sinai: social and family structure. In *Sinai (Human Geography–Part II)*, ed. G. Gvirtzman, A. Shmueli, Y. Gradus, I. Beit-Arieh and M. Har-El Eretz, pp. 807–818, Tel Aviv: Israeli Ministry of Defense Publications.

Palmer, E.H. (1871). *The Desert of Exodus*. Cambridge: Cambridge University Press.

Perevolotzky, A. and Perevolotzky, A. (1979). Agriculture and herding: a traditional food resource of the Gebeliya Tribe. Tel Aviv: The Society for the protection of Nature (in Hebrew).

Revelle, R. (1968). Introduction to historical population studies. *Daedalus* **97**, 353–362.

Robinson, E. (1841). *Biblical Researches in Palestine, Mount Sinai and Arabia Petraea*. London.

Salzano, F.M. (1972). Genetic aspects of the demography of American Indians and Eskimos. In *The Structure of the Human Populations*, ed. G.A. Harrison and A.J. Boyce, pp. 234–251, Oxford: Clarendon Press.

Sharon, M. (1977). The Bedouin and Eretz Israel under Islamic rule. In: *The Desert, Past, Present, Future*, ed. E. Sohar, pp. 199–210. Tel Aviv: Reshafim (in Hebrew).

Spielman, R.S., Neel, J.V. and Li, F.H.F. (1975). Inbreeding estimation from population data: models, procedures and implications. *Genetics* **85**, 355–371.

Stanley, A.P. (1864). *Sinai and Palestine*. London.

Tsafrir, Y. (1970). Monks and monasteries in Southern Sinai. *Qadmoniot* **3**, 2–18.

Wright, S. (1921). Systems of mating. II. The effects of inbreeding on the genetic composition of populations. *Genetics* **6**, 124–143.

Wright, S. (1951). The genetical structure of populations. *Annals of Eugenics* **15**, 323–354.

# 5  *Uncertain disaster: environmental instability, colonial policy, and resilience of East African pastoral systems*

SANDRA GRAY, PAUL LESLIE AND HELEN ALINGA AKOL

## Introduction

In a review paper published in 1980, Rada and Neville Dyson-Hudson observed that there is no universally applicable set of criteria by which all extant nomadic pastoralist societies may be defined, although most have a few basic features in common. These commonalities, that pastoralists keep livestock and tend to move, are sufficiently general to characterize most groups under consideration but fail to distinguish pastoralists from a number of primarily agricultural groups. Nor are efforts to move beyond the general to the specific particularly meaningful for pastoralist societies as a whole. Furthermore, self-identification by pastoralists themselves, although providing insight into historical processes and cultural origins, often fails to correspond to present realities inferred from observed behavior (R. Dyson-Hudson, 1972).

Problems encountered in classifying pastoralist societies derive from their inherent instability. Pastoralism is above all a flexible subsistence strategy, involving opportunistic food production and foraging in addition to livestock exploitation for meat, milk and blood. Although it usually entails mobility of livestock and human populations, in the form of either nomadism or transhumance, it also may incorporate a significant sedentary component. Smith (1992) characterized this variability in pastoralist adaptations as a continuum, along which populations are distributed according to the contribution of livestock production to their total subsistence. This model is most useful for comparative research but should be applied with some caution, because the position assigned to a population along the continuum may vary, depending on the timing of the research and the theoretical bias of the investigator (compare, for example, Dyson-Hudson, 1966, and Wilson, 1985).

99

Regardless of the changeable classifications of pastoralists themselves, there is a striking consistency in representations of the environments in which they live (see, for example, Fratkin, 1991; Majok and Schwabe, 1996; Little *et al.*, 1999*b*). Pastoralist ecosystems are characterized by harsh and unpredictable climates, extreme seasonality, inhospitable terrain, and variable biomass production, and survival is chronically precarious for the species exploiting them. Clinally situated at the interface between higher-productivity ecosystems and tropical or arctic deserts, the boundaries of pastoralist ecosystems are fuzzy, continually shifting in response to climatic flux and changing land-use patterns in adjacent ecozones (Smith, 1992; Stenning, 1959). As a consequence of environmental variability, the position of individual societies along the pastoralist continuum, as well as their location in geographical space and their relationship with neighboring pastoralists, are conditional.

From the beginning of the British colonial era in East Africa, the fluid, ecologically contingent boundaries of pastoralist zones and pastoralist societies were at odds with the more rigid boundaries of colonial polities, defined to conform with colonial spheres of interest (Barber, 1968; Cisternino, 1979; R. Lamphear, 1992; Dyson-Hudson, 1999). These finite political demarcations, however, have persisted into the modern era and have been misconstrued over time to represent the boundaries of meaningful ecological units. Forgetfulness breeds error, and deterioration in environmental quality in such arbitrarily defined systems today tends to be interpreted simplistically as an effect of deviation from ecological equilibrium in a bounded, stable system. This departure is understood moreover to derive from "traditional" pastoralist strategies that lead inevitably to irreversible environmental degradation, not only in pastoralist zones themselves but also in more economically valuable regions abutting them (Hardin, 1968; Fratkin, 1991; Majok and Schwabe, 1996). Among those pastoralist strategies considered to be the most ecologically unsound are unrestricted growth of livestock herds, which increases the risk of overgrazing, and migratory herding, which expands its scope. What is lost in this translation of ecological events is the original tension between indigenous and colonial understanding of boundaries and its contribution to changing ecosystem dynamics.

In this chapter, we examine two pastoralist ecosystems from the perspective of human adaptive strategies in a context of extreme environmental instability, or nonequilibrium. We focus primarily on the relationship between ecosystem fluctuations, livestock population dynamics, and human mobility. We consider the implications of this relationship for

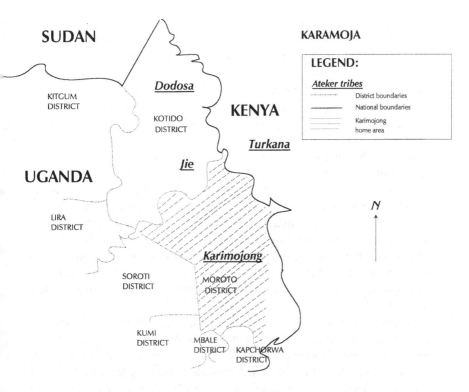

Figure 5.1. Map of Karamoja, Uganda, showing border shared by the Karimojong and Turkana.

political and environmental policy and ecosystem persistence in East Africa; specifically, in reference to two pastoralist populations: the Turkana, of northwest Kenya, and the Karimojong, of northeast Uganda.

### Study populations

Turkana and Karamoja lie at the intersection of modern Kenya, Uganda and Sudan and are contiguous along the Karamoja Escarpment that forms their common border (Figure 5.1). With the neighboring Teso, Jie, and Dodoth people of Uganda, the Toposa and Jiye of Sudan, and the Nyangatom (Dongiro) of Ethiopia, the Turkana and Karimojong comprise the *Ateker* subdivision of the Eastern Nilotic linguistic group (Vossen, 1982; Lamphear, 1992, 1998). The pastoralist populations in Karamoja and

Turkana originated from the same ancestral Nilotic-speaking group (Lamphear, 1976, 1993), and they share many features of their subsistence and social systems. The *Ateker* tribes were the subjects of a number of anthropological studies between the 1940s and 1960s, which provided detailed accounts of pastoralism in northern Uganda and Kenya and in southern Sudan immediately prior to the end of the British era in East Africa (Gulliver, 1955; Thomas, 1965; Dyson-Hudson, 1966; Lamphear, 1976).

In the 1980s and early 1990s, a team of anthropologists and ecologists carried out a long-term study of livestock and human population dynamics, human ecology and human population biology among nomadic Ngisonyoka Turkana pastoralists, in southern Turkana District. Results of these studies are summarized in a recently published synthesis (Little and Leslie, 1999). Consensus of scientists involved in the Ngisonyoka research is that pastoralism in south Turkana is an adaptive subsistence strategy that has shown remarkable persistence in spite of a series of political and environmental catastrophes in the twentieth century (Leslie *et al.*, 1999*b*, p. 373).

In 1998–1999, a field-based study of the effects of economic development and ecological instability on human biobehavioral adaptability was initiated by one of us (SG) among Karimojong agropastoralists in southern Karamoja (Moroto District). Although no other systematic field work has been attempted in Karamoja since the 1950s (Dyson-Hudson, 1966), there have been frequent and disturbing accounts of famine, advancing desertification, disease outbreaks, and widespread armed violence among the Karimojong since the late 1970s – accounts suggesting a breakdown of the responsive subsistence system in place a half-century ago (Gulliver, 1955; Dyson-Hudson, 1966). SG's recent data appear to confirm these reports.

In the following pages, we examine briefly the findings of these two studies. In order to explain the increasing vulnerability of the Karimojong to environmental fluctuations, we focus on the interaction of environmental events and colonial policy in Karamoja in the late nineteenth and twentieth centuries. Why, and how, a similar sequence of historical, political and climatic events in Turkana and Karamoja has produced such profoundly dissimilar results – persistence, in one case, and collapse in the other – in two closely related herding systems is the subject of the concluding sections of the paper.

## Pastoralism and ecosystem dynamics: the case of the Turkana

### *Turkana ecosystem*

Turkana District is a semi-arid scrub savannah, characterized by intense solar radiation, high average daytime temperatures, low rainfall and pronounced seasonality (see Little, 2002). The defining feature of this ecosystem as well as the main challenge to the persistence of resident populations is its profound unpredictability. Annual rainfall averaged just over 300 mm in the decade between 1980 and 1990 (Gray, 1992), but these averages obscure extreme intra- and interannual variability. Convective rains fall in a bimodal pattern across the year in response to movement of the Subtropical High Pressure Area and the Intertropical Convergence Zone (Aryeetey-Attoh, 1997), but their precise location, timing and magnitude are impossible to predict (Figure 5.2). Rainfall also varies in longer cycles of approximately 5–10 years, during which multi-year periods of low rainfall and drought alternate with periods of relatively good conditions (Figure 5.3). Finally, intervals of 20–50 years or more, comprising several of these intermediate-length cycles, may be characterized by chronically low or high rainfall. Regardless of the character of prevailing rainfall patterns, the probability of drought during any single year is high, as is the probability of multi-year drought during any given 5–10 year sequence (Ellis *et al.*, 1987; Little *et al.*, 1999a,b).

In a seminal paper drawing on earlier work by Wiens (1977), Ellis and Swift (1988) argued that extreme fluctuations in rainfall are so frequent in Turkana that interdependent populations are chronically poised between collapse and some stage of recovery. Herd recovery from seasonal droughts is quick, but after multi-year droughts, when livestock losses may exceed 50%, recovery may take upwards of 5 years, under the best of conditions, and far longer, under the worst (Ellis and Swift, 1988; McCabe, 1990a; Scoones, 1993, 1996). In modeling this turbulent environment, the concept of ecological carrying capacity is irrelevant, because no population possibly can attain it (see also Bartels *et al.*, 1993; de Leeuw and Tothill, 1993). Since the maintenance of the size of interdependent populations at or just below carrying capacity is a fundamental feature of ecological equilibrium, Ellis and Swift concluded that ecosystems experiencing fluctuations of the magnitude described for Turkana, where the coefficient of variation in annual rainfall is greater than 33% (see Ellis, 1994), are inherently nonequilibrial. Leslie (1996) later observed that homeostatic or negative feedback relationships between interdependent

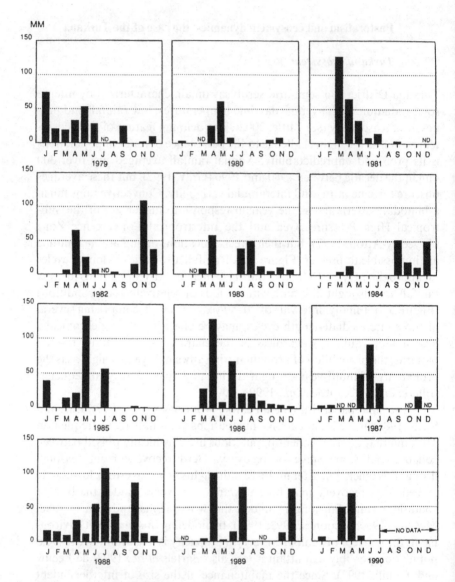

Figure 5.2. Monthly rainfall in south Turkana, 1979–90.

populations, which are critical for the maintenance of ecological equilibrium, are not operational in nonequilibrial ecosystems, because such systems are density-independent, and self-regulatory effects are damped by
the impact of stochastic events.

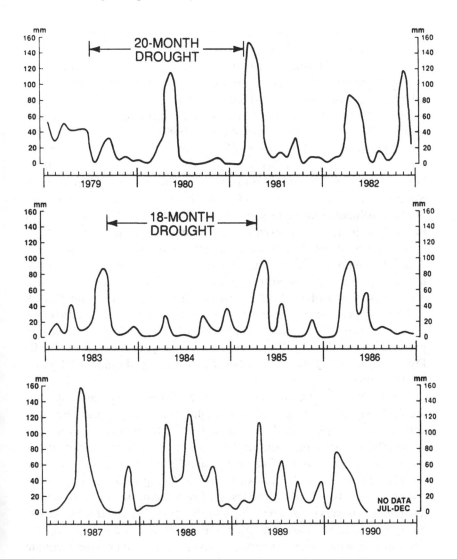

Figure 5.3. Intermediate-length cycles of drought and good rainfall in south Turkana, 1979–90.

### Tracking strategies and Ngisonyoka adaptability

The dissociation of population responses of different species has important implications for human adaptation in nonequilibrial ecosystems. Ngisonyoka pastoralists utilize a number of mechanisms that allow them to track environmental fluctuations, the most essential of which are (1) high stocking rates (N. Dyson-Hudson and Dyson-Hudson, 1982; Scoones, 1993); (2) mixed herding, in which herds include species with different forage needs and of differential resistance to environmental stress (N. Dyson-Hudson and Dyson-Hudson, 1999); and (3) mobility of herders within and across fluctuating territorial and social boundaries (McCabe, 1990b, 1994; Johnson, 1999; McCabe et al., 1999).

Ultimately, the goal of Ngisonyoka tracking strategies is to ensure the persistence of pastoralists themselves in an ecosystem defined by its instability. Fitness of Turkana pastoralists hinges on the maintenance of adequate proportions of humans and livestock over the long term, in the face of extreme fluctuations in the size of herds over the short term. The ratio of livestock to humans in a given homestead must be adequate to meet the nutritional needs of its members, in spite of unpredictably alternating cycles of high and low livestock productivity. Commonly, the ratio of livestock to humans is used to assess the overall resilience of pastoralist systems. A median of 3–4 head of cattle or their equivalents in small stock and camels[1] per capita is the minimum acceptable ratio, while a ratio of six or more livestock units is ideal and is characteristic of the most successful pastoralist societies (N. Dyson-Hudson and Dyson-Hudson, 1982; R. Dyson-Hudson and McCabe, 1985; Fratkin, 1991; Leslie and Dyson-Hudson, 1999).

For the Ngisonyoka, the critical feature of their ecosystem is the unpredictability of this ratio: during drought, livestock numbers plummet, and the ratio drops abruptly to 2 or lower, far below what is needed to sustain the human population. At the same time, extreme variability in the magnitude and duration of environmental "flips" militates against the stabilization of either the human or livestock populations at optimally minimal levels: the smaller a herd is, the lower the probability that any of its animals will survive a severe drought (Scoones, 1993). Herds therefore must be allowed to increase as conditions allow, and the human population must have the capacity to grow commensurately, to provide adequate caretakers for the herds. At the same time, the size of a herder's household should not exceed the nutritional capabilities of his herd during any

---

[1] According to Fratkin (1991) a livestock unit, LU, is equal to one 250-kg cow, 0.8 camels, 11 small stock, or some combination of the three. Leslie and Dyson-Hudson (1999) use a slightly different unit, the standard stock unit (SSU). An SSU is equal to 1 cow, 0.6 camels, or 10 small stock. The relative weights assigned to cows or camels varies, depending on their relative contributions to the species composition of the herds. Recommendations regarding minimal livestock/human ratios also vary according to age, sex and reproductive status of the animals and environmental quality.

environmental cycle. Any sustainable solution to this pastoralist dilemma must facilitate (1) rapid reductions in the human population in response to precipitous livestock losses during bad years and (2) moderate increases in the size of the care taking population during good years. In effect, the success of the Turkana pastoralist system depends on the demographic plasticity of the human population, and demographic adjustments must be less homeostatic (stabilizing) than contingent.

### Ngisonyoka demographic responses

Excessive human mortality is by definition the least adaptive demographic response to worsening environmental conditions, particularly since mortality rates during famine always are highest for young children (Biellik and Henderson, 1981, reported 61% infant mortality and 31% child mortality in relief centers during the 1980–81 famine in Karamoja). There is some indication that disaster-related deaths are minimized in some way in gisonyoka. Indirect estimates of under-5 mortality among African pastoralists range from approximately 12% to 48% (Leslie *et al.*, 1999*a*), and estimates for Turkana nomads are at the low end of this range (Leslie *et al.*, 1999*a*). None of these estimates provide information on periodic changes in mortality levels, although anecdotal evidence suggests that some Ngisonyoka families lost entire cohorts of infants and toddlers to drought and famine or to the epidemics that followed them (Gray, 1992, Pike, personal communication, 1994).

Nonetheless, migration and damping of fertility appear to be more adaptive solutions to the problem of rapid population reduction among Ngisonyoka pastoralists in drought years. The most striking fertility adjustment is a marked seasonality of births, first identified by Leslie and Fry (1989), but there also is a correspondence between intermediate-length environmental fluctuations and fertility rates. In separate studies, Gray (1992) and Leslie *et al.* (1999) reported lower age-specific marital fertility during the drought years of 1979–85 than during the wetter period of 1986–9. Both biological and behavioral factors have been implicated (Gray, 1992; Leslie *et al.*, 1993, 1996, 1999).

Over the long term, migration appears to be the most effective response. During a single-year drought, internal migration may suffice to maintain adequate livestock-to-human ratios by reallocating people and livestock among herding units (Ellis and Swift, 1988; Fratkin, 1991). During multi-year droughts, out-migration is a critical response. In the devastating drought of 1979–81 and the first year of recovery, out-migration increased from approximately 150 persons per 10 000 population, to over 1100 per 10 000; significantly, few emigrants returned to the district (R. Dyson-Hudson and Meekers, 1999; Leslie *et al.*, 1999*b*).

Together, migration and reproductive plasticity have the potential to decrease significantly the size of the human population during drought and the critical years of recovery, at the same time damping mortality effects. Relatively low mortality estimates for nomadic Turkana offer some support for this hypothesis and are suggestive of a population that is well adapted to their stressful environment. Other evidence of the successful adaptation of Turkana pastoralists is their low disease load, overall good health, in spite of limited access to modern health care, and low levels of moderate-to-acute malnutrition (Gray, 1998; Leslie *et al.*, 1999*b*; cf. Shell-Duncan, 1995).

## Colonialism and ecosystem instability: the case of the Karamoja

### *Ecosystem dynamics and Karimojong subsistence strategies*

Karamoja receives substantially higher rainfall than Turkana. We argue nonetheless that Karamoja exhibits characteristics of a nonequilibrial ecosystem. Average annual rainfall in central Matheniko county (Figure 5.4) was 800 mm between 1968 and 1993, and rainfall exceeded 1000 mm in some years (Figure 5.5), but fluctuations in the level, location, and timing of rainfall were as frequent and as variable in magnitude in Karamoja as might be expected in Turkana. Although interannual variability in rainfall in Karamoja was approximately 28% from 1968 to 1993, monthly variability in the same period did not drop below 38% in the rainy season (April through August) and ranged between 70% and nearly 200% in the rest of the year (Figure 5.6).

Extreme variability within annual cycles drives the dual-subsistence strategy of the Karimojong, who traditionally combined nomadic cattle keeping with semi-sedentary agriculture. In good years, rainfall supports sorghum cultivation on a large scale, but in any given year, there is a less than even chance that the harvest will be adequate to support the population through the next growing season (see Figure 5.6). In approximately 1 out of every 3 years, crops are poor or fail completely (Dyson-Hudson, 1966, p.42; see also Campbell *et al.*, 1999). Karimojong persistence in this uncertain agricultural setting depends on livestock keeping and the capacity of the herds to produce adequate milk and blood to sustain the

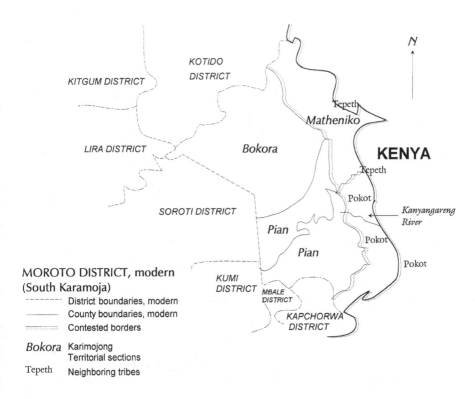

Figure 5.4. Territorial and county boundaries in Moroto District, south Karamoja.

human population during the growing season, in good years, and during the entire year, when crops fail.

Even with this combined subsistence strategy, drought and famine are recurring themes in Karimojong history and are chronicled in the local event calendar (Akol and Gray, unpublished, 1999; see also Lamphear, 1976). In addition to unpredictable annual crop yields, wetter conditions in Karamoja support higher parasite loads in both humans and livestock and more frequent and virulent disease outbreaks. Famine often is preceded by livestock epidemics as well as by drought (see for example Lamphear, 1992, pp. 33–35).

The affiliation of Karimojong families with semi-permanent agricultural settlements, or home areas (*ngireria*), creates the illusion of a sedentary lifestyle. In fact, the Karimojong rely greatly on mobility in response to climatic flux. When the rains are good and crop yields are high, women,

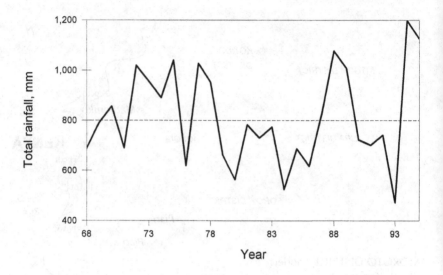

Figure 5.5. Annual rainfall in Moroto District, south Karamoja, 1968–93.

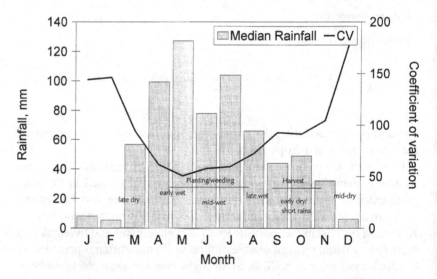

Figure 5.6. Averages and coefficient of variation for monthly rainfall, 1968–93. The figure also shows the correspondence between seasonal rainfall and the agricultural cycle.

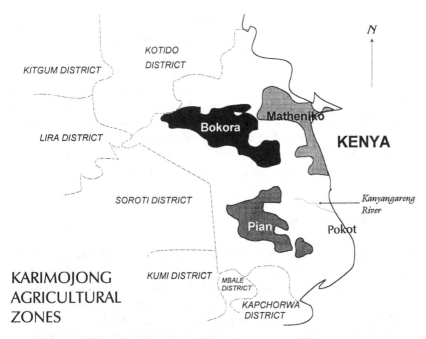

Figure 5.7. Agricultural zones of Pian, Matheniko, and Bokora territorial sections.

young children and the elderly may reside for most of the year in sedentary farming homesteads in the agricultural zones (Figure 5.7), but herding camps are completely nomadic, and their composition is changeable. Livestock are moved in a continuous seasonal pattern that incorporates both large-scale and small-scale migrations. Generally, the population of herding camps includes young girls, boys older than age 5 years, and adult males, who migrate with the herds of cattle, sheep, and goats between the central farming zone and western and southern grazing areas. In fact, there is a constant flow of people, information, and livestock between cattle camps and agricultural settlements, and it is a key mechanism to improve the quality of information about water and forage availability. This exchange also ensures that milking animals and a supply of fresh milk are available for young children remaining at the homesteads, and it serves as an effective warning system, alerting members in both locations in the event of raids, government operations, disease outbreaks, deaths, and labor shortages (which may affect either herding or agriculture).

In years of poor rainfall, the entire homestead moves with the herds (SG, personal observation, 1998–99), with the exception of the oldest individuals who are too weak or too ill to travel. Epidemics also cause people to abandon settlements for sparsely populated, mobile livestock camps. In the past, drought triggered mass migrations of entire clans to other parts of Karamoja as well as to Teso and Mbale, in the west and south, and to Turkana, in the east. Informants frequently reported moving from area to area for a decade or more before environmental conditions enabled them to return to their traditional home area. Internal and external migration, the scale of which is determined by prevailing environmental conditions, thus has been as critical a response of the Karimojong to their unstable environment as it is for nomadic Turkana.

### Karimojong adaptability

Although many features of Karimojong agropastoral subsistence still are in place, SG's recent research points to the intensification of selection pressure in this population in recent decades. Preliminary analysis of reproductive histories from over 300 Karimojong women, plus their mothers and their co-wives, suggest decreased marital fertility since the 1950s, which is linked as well to higher infant and toddler mortality (Gray and Akol, 2000). Anthropometric and nutritional examinations conducted in 1998–99 among nearly 500 Karimojong children, ranging in developmental age from early infancy through puberty, indicate that stunting and wasting are prevalent, while the timing of puberty appears to be in the late teens, among females, and older, among males. A frequently cited cause of child deaths is *"lobulbul"*, or "swelling", which may translate into kwashiorkor.

Endemic disease and chronic illness, rather than food scarcity, appear to be the proximal determinants of widespread child malnutrition and developmental delays (there were favorable rains and good harvests from 1994 to 1998). Malaria, gastrointestinal tract infections, tuberculosis, neonatal tetanus, and a variety of skin infections and other parasitic diseases are rampant. Frequently reported secondary sterility may point to prevalent sexually transmitted diseases. Interestingly, the Karimojong have much enhanced access to formal health care, relative to the Turkana. Childhood immunization coverage, in particular, has been improved in Karamoja through the introduction of mobile clinics. In Bokora subsection, most mothers sampled were able to provide confirmation (immunization cards) of DPT (diphtheria, pertussis, tetanus), measles, and polio vaccinations for their children aged 2–3 years and younger.

Symptoms of social and intra-familial stress are endemic alcoholism and domestic violence. Suicide is a frequently reported cause of death. Automatic weapons are widely available, and the Karimojong are very much an armed society; shootings and murder are routine at trading centers and administrative posts and are not uncommon within homesteads. Armed attacks are reported throughout the year, but are most frequent during cattle migrations, at which time, travel along any of the major roads or tracks in the district, either on foot or in a motor vehicle, is especially dangerous.

Political interactions are almost universally armed interactions. Inter-tribal raiding – a longstanding, controversial solution to the problem of cattle losses among East African pastoralists – has increased recently among the Karimojong, Pokot (Suk, Upe), and Tepeth. Traditionally there were strong sanctions against intra-tribal violence among the Karimojong (Dyson-Hudson, 1966; Lamphear, 1992), but deadly Karimojong-on-Karimojong raiding became widespread during the 1980s and early 1990s and erupted on a district-wide scale between 1996 and 1999, in spite of both government and nongovernmental organization (NGO) peace initiatives (GOU, 1994; SG, personal observation, 1996, 1998–99. See also Little *et al.*, 1999*a*). Intra-societal killing, even outside of the context of cattle raids, has emerged as the interaction of choice among the three territorial sections of Pian, Bokora, and Matheniko (see Figure 5.4).

From all perspectives, the current status of the Karimojong stands in marked contrast to their situation at mid-century (Jelliffe *et al.*, 1964; Dyson-Hudson, 1966) and to what we have observed among Ngisonyoka pastoralists. Yet both Karamoja and Turkana experienced severe droughts, epidemics, and intensification of inter-tribal raiding between the late 1960s and early 1980s. The Turkana pastoralist system had achieved a remarkable recovery by the end of the 1980s (McCabe, 1987; Galvin, 1992), while the Karimojong continued to experience steadily deteriorating environmental and economic conditions into the 1990s. Emergency food relief has been distributed in Karamoja in nearly every year since 1980 (St Kizito Hospital, 1994; SG, personal observation, 1996, 1999). According to the few reports available, in the years since 1979, the Karimojong have suffered almost total cultural and ecological collapse (Alnwick, 1985; Dyson-Hudson, 1989), while Ngisonyoka pastoralists and their drier ecosystem have experienced more characteristic cycles of drought and recovery (Ellis *et al.*, 1987; Leslie *et al.*, 1999*b*). We propose that the crisis in Karamoja is primarily ecological, and that it is rooted in early colonial policies toward pastoralists in northern Kenya and Uganda. While this crisis was exacerbated by civil war in Uganda after 1969, it had its beginnings in an altered relationship between the Karimojong and their environment, their herds, and their neighbors. This change derived directly from colonial policies aimed at fixing ecosystem boundaries and damping Karimojong mobility.

### Colonial policy in Karamoja, 1911–54

British involvement in Karamoja dates from the Anglo-German agreement of 1890, which divided the spoils in colonial East Africa. For the British, the objective not only was to establish which of the most productive sections of East Africa fell within their sphere of influence but also to annex a permanent buffer zone between the Ethiopian empire and areas of British settlement in the fertile highlands of British East Africa (modern Kenya) and in Buganda (southern Uganda). This buffer zone encompassed a large section of the vast savannah of eastern Africa, where specialized pastoral subsistence systems had evolved over more than two millennia (Smith, 1992). From the outset, neither the British Foreign Office nor the colonial administration had any interest in administering this remote, inhospitable region and its unruly populations, and involvement in the 1890s and early 1900s was limited to a few poorly supported military expeditions (Barber, 1968; Lamphear, 1992).

A profitable ivory trade in Karamoja, developed in this period by Swahili, Ethiopian, and European traders, ultimately would play a critical role in the development of colonial policy toward the district. The ivory traffic provided a means for the opportunistic Karimojong to rebuild livestock herds that had been destroyed by rinderpest and bovine pleurapneumonia epidemics in the 1890s. It also provided an abundant supply of guns with which to mount cattle raids on neighboring pastoralists (Barber, 1968). By 1910, Karimojong livestock herds had returned to and probably had exceeded their pre-rinderpest levels, and herders again were expanding the range of their migrations into grazing areas on the perimeter of their territorial sections, areas that had been emptied, temporarily, by livestock disease.

Their expanded migrations once again brought the Karimojong and neighboring pastoralists into close contact along the changeable borders of their outlying rangelands, and inter-tribal friction in contested zones intensified. A widening periphery also brought native livestock closer to white settlement zones, increasing the risk of exposure of newly introduced European herds to indigenous livestock diseases and epidemics. Escalating inter-tribal violence, heightened risks of the spread of livestock disease, and encroachment of Ethiopian ivory traders on the buffer zone induced the British to establish a military administration in Karamoja in 1911 (this was transferred to civilian rule in 1924). Expansion of effective colonial influence into the pastoralist zone had three long-lasting consequences for the structure of Karimojong pastoralism. First, the establishment and closure of political boundaries between British East Africa and Uganda forced pastoralists on both sides of the border to alter as well as to curtail their migratory behavior. A second effect of colonial administration was the disproportionate redistribution of livestock among neighboring pastoralists (Barber, 1968; Lamphear, 1992). The third and most enduring legacy of British administration has been a series of ecological catastrophes from 1967 to the present, which can be linked directly to the first two effects and to a belated and misinformed economic development program at the close of the colonial era.

*Borders and grazing rights*

In 1902, the northern border between the Ugandan Protectorate and British East Africa was formalized along the length of the Turkwel River, from Lake Turkana south. As specified by the Outlying Districts Ordinance of the same year, the districts on either side of this boundary, namely, Turkana and Karamoja, were designated "closed districts" (Barber, 1968; Cleave, 1996; Fleay, 1996), which placed them off limits to traders, missionaries, and all private expeditions (Lamphear, 1992). Because the two districts were administered as parts of two distinct colonial polities, movement of pastoralists and herds across their common border also was restricted. In the north, the border dissected the Turkana tribe and in the south, the Pokot.

Since the 1890s, the border between the two protectorates had fluctuated east and west along different lines drawn through the valley of the Turkwel and Kanyangareng Rivers (Figures 5.4 and 5.7). In his 1968 analysis (p. 161), Barber observed that the upper Turkwel River valley had for years brought Pokot and Karimojong herders together in their quest for dry season water and forage. The boundary between them was fluid, because it was contingent on environmental conditions and the status of their livestock herds. The colonial border formalized their separation and arbitrarily assigned permanent territorial rights based on whichever tribe seemed to be predominant in the valley at the time the boundary was set. Pokot currently occupying the grazing areas west of the new border, in Karamoja, who now were cut off from their territories to the east, were granted permanent grazing rights to the contested Turkwel and Kanyangareng valleys. Pian territorial section of the Karimojong was denied access to the same grazing areas, which they had dominated a decade earlier.

## Livestock and grazing rights

From the beginning of their military administration of Karamoja and Turkana, the British carried out numerous punitive expeditions, confiscating large numbers of livestock as a penalty for raiding, rebellion, and refusal to pay hut tax. Impounded livestock often were redistributed among victims of the raids – and tribes who were generally more receptive to British rule. British administrators accepted the victims' estimates of their losses, which often far exceeded their losses in fact (Lamphear, 1992). Most frequently, the Pokot were the recipients of this compensatory bounty, while the Karimojong and the Turkana in particular sustained heavy losses (Lamphear, 1992. See also Cleave, 1996).

A disparity thus was created between the size of Pokot herds and the size of their rangelands, causing them to push farther west into Karamoja to occupy additional Pian grazing areas as well as sections of southern Turkana. Indeed, during this period, the Pokot, with the support of the colonial government, are reported to have experienced greater territorial expansion than any of the other pastoral tribes in the region, largely at the expense of the Turkana and Karimojong. To redress Pian losses, the Ugandan administration resettled them permanently farther north in Karamoja, in the territories of Bokora and Matheniko territorial sections (Barber, 1968).

As a result of these policies, mobility of people and herds in response to changing environmental conditions – a crucial strategy to reduce grazing pressure on the rangelands and to damp the effects of drought and disease – increasingly was restricted. Furthermore, indigenous mechanisms that had assured continual adjustments in the size, location, and distribution of tribal livestock herds – the equivalent of a pastoralist balance of power (see Spencer, 1998) – were deactivated. Over the next four decades, outbreaks of hostility among the Pokot, Karimojong, and Turkana over contested territories along the rigid Kenyan/Ugandan border would increase steadily in frequency and severity as ecosystem stress intensified.

### Ecosystem instability and economic development initiatives in Karamoja

*Economic interventions in Karamoja, 1924–62*

In the early years of civilian administration in Karamoja, there had been a passing interest in agricultural development and permanent settlement of the Karimojong in farming villages. Even then, these efforts were thwarted by the district's unpredictable climate. In subsequent years, there were tentative efforts to develop a cattle trade to supply beef to markets in Kenya and Europe, but there was a transparent lack of commitment to these programs on the part of the colonial administration in Entebbe. In fact no veterinary officer was appointed for Karamoja until 1948, and no more than one such officer was appointed for any period thereafter (Barber, 1968, p. 211). In 1924, it was clear to the government that economic development of Karamoja was beyond its fiscal means, and the then Governor of Uganda issued a decision that the best policy in relation to the district was to allow the people to continue to practice traditional pastoralism (Barber, 1968, p. 209).

By 1952, escalating inter-tribal hostility among Karimojong Pian and Pokot, made increasingly deadly by ease of access to guns, represented a substantive threat to the stability of the nascent Ugandan republic. Deteriorating political relations in the north now were interpreted by colonial policy-makers as an interrelated effect of underdevelopment, isolation, and encroaching desertification. All of these were perceived to be direct outcomes of subsistence pastoralism, a view that conveniently ignored the altered configuration of land, livestock, and people. In 1954, the colonial government reversed its 1924 policy and launched an accelerated program of agrarian reform in the district, the objectives of which were, first, to check soil erosion and desertification by reducing the size of the herds, and second, to transform the Karimojong into 'modern' ranchers and farmers who could participate in some meaningful way in Ugandan economy and society (Barber, 1968; Fleay, 1996). The means to these two ends were the same: de-stocking, enclosure of grazing lands, water development, expansion of veterinary care, continuation of Karamoja's "closed" status, agricultural development and finally, permanent settlement of the Karimojong on agricultural and ranching schemes (Barber, 1968, p. 215; Fleay, 1996). The effect of this intervention was in fact to accelerate the rate of overgrazing and desertification, to aggravate the contraction of tribal grazing areas, and to inflame further both inter- and intra-tribal tensions.

As independence loomed, the Karamoja crisis remained a primary focus of transition teams. A series of investigative commissions in 1961 re-affirmed that rapid development of Karamoja was the best means to contain and eventually eliminate unrest in the pastoralist north (Fleay, 1996). In the words of the Karamoja Security Committee report of 1961, "If Karamoja is to cease to be the problem that it is now the pace of development must be forced and forced hard" (Barber, 1968, p. 219. But see Fleay, 1996). "Development" by now had become synonymous with permanent settlement and agriculture, because settled people were easier to monitor and control, and because settlement was perceived to be a sound solution to the problem of overgrazing and environmental degradation (see also Fratkin, 1991; Majok and Schwabe, 1996). Settlement became a keystone of development policy.

*Ecosystem dynamics, 1924–79*

Between 1924 and 1954, the Karimojong were allowed to continue to pursue their agropastoralist subsistence. Unfortunately for later events in Karamoja, the 1924 policy recommendation did not advocate restoration of pastoralists' mobility and autonomy, which had been undermined seriously by the closure and shrinkage of the district. The situation was quite the reverse, in fact, for by this time, large sections of Matheniko and Bokora grazing lands had been set aside as forest and wildlife reserves and were now also off-limits to pastoralists (D.O.S., 1972). It is perhaps the cruellest of ironies that the next 30 years were a period of unusually favorable conditions for herd growth, and both the human and livestock populations increased dramatically (Cisternino, 1979). Although the precise magnitude of the increase cannot be estimated reliably, it is certain that sustained population growth played a major role in the events that followed (Barber, 1968, p. 217; Cisternino, 1984; Alnwick, 1985). Rapid herd growth, in combination with hampered mobility and shrinking rangelands, triggered increased competition for remaining "open" water and grazing resources. In the 1950s, Neville Dyson-Hudson, who conducted his fieldwork in this period, suggested that raiding might be reduced by restoring traditional territorial flexibility, but his recommendation was not compatible with current policy (Cleave, 1996. But see Fleay, 1996).

As we have summarized above, the solution adopted by the transitional government was to undertake rapid economic development in the district and to enforce environmental restrictions with renewed vigor. Furthermore, development policy in this period increasingly was predicated on an assumption of environmental equilibrium and the need to restore carrying capacity and to stabilize the livestock population (see Fratkin, 1991; Majok and Schwabe, 1996). Proposed changes in livestock and rangeland management based on the notion of carrying capacity may have seemed reasonable to outsiders whose entire experience of the Karimojong pastoralism was limited to a period of uncommonly favorable conditions in an artificially bounded system, but these proposals ran counter to everything the Karimojong themselves had learned about anticipating and negotiating abrupt change in the quality of their environment. Other development initiatives, however, such as expanded veterinary services and extensive borehole and dam construction, were readily accepted by opportunistic pastoralists, because they helped to sustain herd growth during a drought-free cycle. Boreholes were embraced with considerable enthusiasm, given the restrictions on pastoralists' ability to exploit water resources throughout the district. Selective acceptance of development initiatives by pastoralists resulted in prolonged concentration of herds around boreholes, which aggravated overgrazing, and in improvements in livestock health, which enhanced herd survival and growth.

Ecological conditions shifted abruptly with the arrival of the drought of 1967–68, a rinderpest epidemic in 1968–69, and a cholera outbreak in 1971 (Akol and Gray, Local event calendar for Bokora and Matheniko Karimojong. Unpublished, 1999). A second multi-year drought occurred in 1974–75 and a third, in 1979–81. In the drought of 1979, the Karimojong livestock herds "crashed", the crops failed, and the following year brought famine to Karamoja. The size of the herds at the onset of the 1970s and the scale of overexploitation of remaining rangelands appear to have been critical factors in the inability of the ecosystem to attain even partial recovery between droughts. Concomitant increases in the human population also have been implicated in the horrific famine of 1980 (Cisternino, 1984; Alnwick, 1985).

### Politics and instability

The impact of environmental events in Karamoja was heightened by increasing politicization of access to steadily deceasing grazing resources. By 1967, Pian displaced by Pokot encroachment on their grazing lands had been resettled more or less permanently on the border between Matheniko and Bokora territory. The Matheniko range already had been reduced by the establishment of a forest reserve in the northern part of their territory. As recurring droughts pushed Matheniko farther from their central zone, they came into conflict with resettled Pian as well as with migrating

Bokora. Matheniko subsection began to mount raids against the Pian and Bokora as early as 1974–75. Meanwhile, Amin's ouster in 1979 resulted in a flood of automatic weapons into Matheniko County, enabling younger men, whose access to livestock traditionally had been limited by the elders and the generation-set system (Lamphear, 1998), to organize the equivalent of armed militias, which they used to mount intra-tribal raids and amass large herds. Authority and power now were transferred to those who controlled the guns and the herds.

With the addition of this new dimension in political relations, distribution of livestock among the Karimojong population became increasingly inequitable (Gertzel, 1988), representing a striking departure from the apportionment of the herds 30 years earlier (Gulliver, 1955; Dyson-Hudson, 1966). By the mid–1980s, there existed in Karamoja a large, impoverished underclass with virtually no access to the pastoral sector, who had become chronically vulnerable to food shortages and dependent on international donors for economic assistance and food aid.

Beginning in the drought of 1967–68, missionaries, the United Nations, government agencies, and NGOs (Barber, 1968; M. Cisternino, personal communication, 1996; SG, personal observation, 1999) became a constant presence in Karamoja. The impact of drought and famine in the 1970s and 1980s was damped somewhat by their interventions, in the form of food-for-work programs, economic assistance, famine relief on a grand scale, and the expansion of healthcare facilities. At the same time, the international presence helped to politicize the Karamoja crisis.

By the early 1980s, Bokora and Pian territorial sections had suffered devastating losses of livestock as well as grazing areas to armed Matheniko raiders. Cattle losses to Matheniko raids had precipitated famine in Bokora and large-scale out-migration of the Bokora population in 1974–75 (Akol and Gray, unpublished, 1999). Impoverished pastoralists began to take advantage of the economic and educational opportunities offered by external donors and subsequently were able to recoup their losses and acquire considerable wealth and influence through their integration into the international communities. Today, a majority of the educated elite in Karamoja, who control access of Karimojong to government programs and international donors, are Bokora or Pian, while Matheniko have become increasingly marginalized. Bokora and Pian involvement in modernization and national politics and their rumored backing of raids against the Matheniko have increased the opposition of the latter to external governance and to economic development. At the present time, government and NGO workers alike perceive the Matheniko to be irrationally and violently committed to a way of life whose passing is considered to be a

forgone conclusion, while Bokora and Pian are openly receptive to economic development and change. The precarious socioeconomic and political position of the Matheniko has been worsened in recent years by the
formation of an alliance of Tepeth and Pokot, who continually harry
Matheniko herding camps and settlements along Moroto's eastern border.

## Conclusions: interpreting instability

We summarize this complex sequence of cause-and-effect as follows: first,
the political closure of Karamoja in the first half of the twentieth century,
which derived from a radical divergence of indigenous and colonial understanding of ecosystem behavior and the nature of boundaries in inherently
unstable systems, undermined the environmental and social resilience of
the Karimojong. Contraction of rangelands, reconfiguration and formaliz-
ation of tribal boundaries, reallocation of herds, and containment of
pastoralists' mobility, coinciding with a prolonged period of favorable
rainfall and rapid growth of the livestock herds, resulted in environmental
degradation on an alarming scale by the 1950s. This ecological effect
conformed to predictions of ecosystem behavior based on the assumptions
of popular equilibrium models. Development projects subsequently under-
taken to restore equilibrium in Karamoja succeeded only in exacerbating
the environmental effects of more than a decade of arid conditions and
drought. Fortuitous acquisition of large quantities of automatic weapons
by Matheniko warriors abruptly changed the nature of growing competition for dwindling resources and created gross inequities in the agropastoral system. Raiding, unrelenting drought, and recurring famine and
international intervention created a differentially vulnerable and increasingly politicized and violent society. With the exception of a brief interlude
between 1994 and 1996, the cycle of drought, disaster, and violence has
been unbroken since the late 1970s.

The crisis in Karamoja thus is at once ecological and political. On the
surface, it is an effect of a lack of concordance between colonial policy and
later development initiatives (Barber, 1968). Ultimately, it is rooted in a
superficial and deeply flawed European understanding of African savannah ecosystems and the nature of environmental instability in Karamoja.
At the turn of the nineteenth century, the Karimojong were a highly
mobile component of a profoundly unstable ecosystem, and neither of
these ecological units was compatible with the closed and immobile political structure imposed upon it by the colonial administration. Ultimately,
both the ecosystem and the human system collapsed, and neither appears

to be able to sustain a recovery. The human population now appears to be far more vulnerable to environmental vicissitudes than was true 50 years ago.

In a far less productive and more unstable ecosystem, Ngisonyoka pastoralists were able to recover from colonial policies that were equally restrictive (Lamphear, 1992). Indeed, the livestock losses suffered by the Turkana during colonial punitive campaigns were of a scale larger than that incurred by any other pastoralist group in East Africa (Lamphear, 1992). Once pacified, however, they also were left alone to resolve emerging inequities. Aside from relief efforts to deflect impending famine, which in fact were required fairly frequently after the confiscation of the herds, Turkana also remained isolated, undeveloped, and closed. Why such neglect should have altered one system so utterly, while facilitating the recovery of another is of more than academic interest, because it has bearing on the continued failure of economic development in Africa's pastoralist zones and present impoverishment and dependency of pastoralists.

A series of recent volumes presents a monotonous litany of failed economic development projects in Africa that have left both the environment and pastoralists far less resilient than they were prior to development (Behnke and Scoones, 1993; Scoones, 1994; Majok and Schwabe, 1996. See also Fratkin, 1991). In our view, the ecological and economic crises prevailing in so many of Africa's pastoralist lands have been brought about by widespread application of a model with inadequate time-depth and by little or less attention to the interaction of historical and environmental processes.

Karamoja's deceptively high average rainfall – combined with its high visibility in NGO, international relief agency, and missionary circles – has been the stimulus for substantial government and NGO backing of agricultural development since the 1980 famine. None of these projects have been sustainable. Crop failures, food scarcity, and famine have been reported with increasing frequency in the 1980s and 1990s (Akol and Gray, unpublished, 1999; H.A. Akol, personal communication, July 1999; Alnwick, 1985; SG, personal observation, 1994, 1998), even with substantial external support. In spite of these failures, pastoralism is overtly discredited and blamed for the continuing poverty and violence in the region, and traditional territorial affiliations are perceived to be redolent of tribalism and barbarism.

Development initiatives in Karamoja point to what may be the critical difference between Turkana and Karamoja: unlike Karamoja, Turkanaland permits no illusions of economic potential. It was, and remains, essentially a desert. No systematic economic development of the district was attempted by either the colonial or Kenyan governments, although Turkana was the target of economic interventions by the Norwegian Agency for Development during the 1970s and 1980s (Fratkin, 1991; Dyson-Hudson, 1999). These projects differed from development schemes in Karamoja in important ways: first, they provided alternatives to pastoralism but were not aimed explicitly at eliminating the indigenous pastoralist system. Second, by providing economic assistance, they buffered the Turkana somewhat from the effects of the terrible droughts of the 1970s and early 1980s but also provided them with means to acquire livestock and return to the pastoral sector when the droughts were over. Finally, a major contribution was the construction of a tarmac road that runs the length of Turkana district, from north to south (R. Dyson-Hudson, 1999). The road has opened the district to travel and trade, and barter of livestock for sugar, maizemeal, tea, tobacco and other goods now has become an important component of pastoralists' subsistence. Interestingly, the road also may enhance the effectiveness of established mechanisms that track drought and recovery, by overtly welcoming out-migration of dissatisfied and failed pastoralists in a way that Karamoja's impassable, nearly invisible tracks do not (there is substantial political opposition to the paving of the Karamoja's main trunk road).

Differential experience of economic development after the colonial era is only one of the probable consequences of distinct ecosystem dynamics that may have bearing on the responses of the Turkana and Karimojong to similar political and environmental events. Migration in Turkana is qualitatively and quantitatively different from that in Karamoja: Turkana move smaller units over less prescribed routes. The distribution of herding units matches the patchy, unpredictable distribution of water and grazing resources. In Karamoja, on the other hand, where vegetation tends to be at once more dense but more strictly seasonal in distribution, where the torrential rains of the wet season and the fierce sandstorms of the dry season constrain movement for prolonged periods, and where swamps and black cotton soil obstruct cross-country travel, it is not unusual to see tens of thousands of head of cattle moving *en masse* along established routes, to particular grazing areas, and at specified times during the annual cycle. In the final analysis, closure of southern Turkana may have had a less negative impact on the pastoralist system, because seasonal movements of herds are less patterned, and the ecosystem itself is more sparse and less predictable.

A crucial difference is the magnitude of intra-tribal conflict and an apparent lack of social cohesion among the Karimojong, relative to the

Turkana. The heightened risk of armed assault hampers access to both pastoral resources and to district infrastructure; it is probably the primary distal determinant of poor health and nutrition and high mortality of Karimojong women and children (Devlin, 1998). Raiding and banditry, however, are long-standing responses of *Ateker* pastoralists to drought and cattle losses, and there is a an inverse relationship between worsening environmental conditions and the intensity of inter- and intra-tribal hostility in these societies (Lamphear, 1998; Little *et al.*, 1999a). Karimojong-on-Karimojong violence appears to be of a different species altogether. Outbreaks of banditry occurred among the Turkana in the 1970s and 1980s (R. Dyson-Hudson, 1999; Little *et al.*, 1999a), but they never took the form of the institutionalized murder that has emerged in Karamoja. Social disparities and conflict are inherent to generation-set systems in East Africa (Simonse and Kurimoto, 1998; Spencer, 1998), but the Turkana system appears to have facilitated resolution of internal discontent in a context of decreasing resources, whereas the ecological sequence in Karamoja only exacerbated inter-generational friction (Lamphear, 1998; Leslie *et al.*, 1999b, p. 369).

To what degree is the quality of Karimojong violence a legacy of the Ugandan civil war and a series of brutal Ugandan regimes? Access to automatic weapons after 1979 is most certainly a critical factor in the Karimojong equation. The pillage of Amin's armory in Moroto by Matheniko warriors in 1979 ensured the access of young men to livestock, and through the barter of livestock, to ammunition and more guns via the arms trade in Sudan. Furthermore, a campaign to disarm the Turkana was carried out successfully by the Kenyan government in the 1970s, whereas several such campaigns in Karamoja had only a temporary impact, and their savagery further incensed the Karimojong population (Akol and Gray, unpublished, 1999). While civil war raged in western and southern Uganda, the Karimojong evolved into an armed society, capable of repelling both insurgents from outside the district and government troops alike. In their adoption of the culture of the gun, the Karimojong themselves were instrumental in maintaining the status of Karamoja as a closed system.

To argue that the transformation of Karimojong society and subsistence and the persistence of Turkana pastoralism have their origins in ecological instability is not to ignore the impact of these other sociopolitical factors. Today, the Karimojong use and trade automatic weapons as part of their daily social interactions; the Turkana do not (but see Little *et al.*, 1999a). Political instability in Uganda after 1962 played a major role in the emergence of the Karimojong arms trade, and the arms trade has helped to maintain the politicized structure that emerged in Karamoja after the 1970s. Be that as it may, both politics and weapons there are linked to cattle, and cattle still are linked inextricably to ecosystem dynamics.

We conclude this analysis with a caveat that pertains to the study itself of pastoralists and brings us back to our initial observation that pastoralists defy classification. The adoption of a nonequilibrial model to explain pastoralist ecosystem dynamics does not alter this reality, but it does illuminate it. Pastoralism *must* be a highly flexible subsistence strategy, for the simple reason that environmental instability constrains human adaptability in different ways in different ecosystems, and in different ways *at different times* in the same ecosystem: this is perhaps the main fruit to be gleaned from the Karimojong–colonial interaction. By implication, our preferred model may be effective in explaining the past behavior of humans in response to instability in a given ecological and temporal setting; it may have at best a limited capacity to predict human behavior in the same unstable ecological setting at some future time. Our acknowledgment of its limitations certainly does not detract from the model's explanatory power in relation to past sequences. In pastoralists' understanding, however, to know the past is to anticipate a future that is absolutely unpredictable, because it comprises a dynamic set of interacting social and environmental possibilities with indeterminate probabilities. And that, after all, may be the best definition of pastoralism we can offer.

### Acknowledgments

This research was funded by the General Research Fund of the University of Kansas, Wenner–Gren Foundation Grant 6276, and National Geographic Society Grant 6048–97.

### References

Alnwick, D.J. (1981). *Towards a Food Aid Policy for Karamoja*. A report to UNICEF, June 2, 1981. Kampala, Uganda: Karamoja Emergency Relief Program, UNICEF.

Alnwick, D.J. (1985). Background to the Karamoja famine. In *Crisis in Uganda: The Breakdown of Health Services*, ed. C.P. Dodge and P.D. Wiebe, pp. 127–141. Oxford: Pergamon Press.

Aryeetey-Attoh, S., ed. (1997). *Geography of Sub-Saharan Africa*. Saddle River, NJ: Prentice-Hall.

Barber, J. (1968). *Imperial Frontier*. Nairobi: East African Publishing House.

Bartels, G.B., Norton, B.E. and Perrier, G.K. (1993). An examination of the carrying capacity concept. In *Range Ecology at Disequilibrium: New Models of Natural Variability and Pastoral Adaptation in African Savannas*, ed. R.H. Behnke Jr., I. Scoones & C. Kerven, pp. 89–103. London: Overseas Development Institute.

Behnke, R.H. Jr. & Scoones, I. 1993, Rethinking rangeland ecology: Implications for rangeland management in Africa. In *Range Ecology at Disequilibrium: New Models of Natural Variability and Pastoral Adaptation in African Savannas*, ed. R.H. Behnke Jr., I. Scoones and C. Kerven, pp. 1–30. London: Overseas Development Institute.

Biellik, R.J. and Henderson, P.L. (1981). Mortality, nutritional status, and diet during the famine in Karamoja, Uganda, 1980. *Lancet*, **II**, 1330–1333.

Campbell, B.C., Leslie, P.W., Little, M.A., Brainard, J.M. and DeLuca, M.A. (1999). Settled Turkana. In *Turkana Herders of the Dry Savanna. Ecology and Biobehavioural Response of Nomads to an Uncertain Environment*, ed. M.A. Little and P.W Leslie, pp. 333–352. Oxford: Oxford University Press.

Cisternino, M. (1979). Karamoja: The Human Zoo. The History of Planning for Karamoja with some Tentative Conterplanning [sic]. Ph.D. thesis, Post-Graduate School for Development Studies, Swansea University, UK.

Cisternino, M. (1984). *Karamoja: Patterns of Traditional Work versus Modern Sector Employment. Part One: Introductory Notes*. Moroto Diocese Social Services & Development, Uganda. Socio-economic Evaluation of Crash Labor Employment Programmes in Karamoja, Geneva: ILO.

Cleave, J. (1996). First posting. In *Looking back at the Ugandan Protectorate: Recollections of District Officers*, ed. D. and M.V. Brown, pp. 30–38. Dalkeith, Western Australia: Douglas Brown.

de Leeuw, P.N. and Tothill, J.C. (1993). The concept of rangeland carrying capacity in Sub-Saharan Africa: Myth or reality? In *Range Ecology at Disequilibrium: New Models of Natural Variability and Pastoral Adaptation in African Savannas*, ed. R.H. Behnke Jr., I. Scoones and C. Kerven, pp. 77–88. London: Overseas Development Institute.

Devlin, H. (1998). *Patterns of Morbidity in Karamoja, Uganda, 1992–1996*. M.A. thesis in Anthropology, University of Kansas.

D.O.S., Directorate of Overseas Surveys for the Uganda Government (1972). D.O.S. 426 (Series Y732). Reprinted by U.N.D.P., 1992. Kampala: Government of Uganda.

Dyson-Hudson, N. (1966). *Karimojong Politics*. Oxford: Clarendon Press.

Dyson-Hudson, N. and Dyson-Hudson, R. (1982). The structure of East African herds and the future of East African herders. *Development and Change* **13**, 213–238.

Dyson-Hudson, N. and Dyson-Hudson, R. (1999). The social organization of resource exploitation. In *Turkana Herders of the Dry Savanna. Ecology and Biobehavioural Response of Nomads to an Uncertain Environment*, ed. M.A. Little and P.W Leslie, pp. 69–86. Oxford: Oxford University Press.

Dyson-Hudson, R. (1972). Pastoralism: self image and behavioral reality. *Journal of Asian and African Studies* **7**, 319–324.

Dyson-Hudson, R. (1989). Ecological influences on systems of food production and social organization of South Turkana pastoralists. In *Comparative Socioecology: The Behavioral Ecology of Humans and Other Mammals*, ed. V. Standen and R.A. Foley, pp. 165–193. Oxford: Blackwell.

Dyson-Hudson, R. (1999). Turkana in time perspective. In *Turkana Herders of the Dry Savanna. Ecology and Biobehavioural Response of Nomads to an Uncertain Environment*, ed. M.A. Little and P.W Leslie, pp. 25–40. Oxford: Oxford University Press.

Dyson-Hudson, R. and Dyson-Hudson, N. (1980). Nomadic pastoralism. *Annual Review in Anthropology* 9, 15–61.

Dyson-Hudson, R. and McCabe, J.T. (1985). *South Turkana Nomadism: Coping with an Unpredictably Varying Environment*. New Haven, CT: Human Relations Area Files, Inc.

Dyson-Hudson, R. and Meekers, D. (1999). Migration across ecosystem boundaries. In *Turkana Herders of the Dry Savanna. Ecology and Biobehavioural Response of Nomads to an Uncertain Environment*, ed. M.A. Little and P.W Leslie, pp. 303–313. Oxford: Oxford University Press.

Ellis, J.E. (1994). Climate variability and complex ecosystem dynamics: implications for pastoral development. In *Living with Uncertainty: New Directions in Pastoral Development in Africa*, ed. I. Scoones, pp. 37–46. London: Intermediate Technology Publications.

Ellis, J.E. and Swift, D.M. (1988). Stability of African pastoral ecosystems: alternate paradigms and implications for development. *Journal of Range Management* 41, 450–459.

Ellis, J.E., Galvin, K., McCabe, J.T. and Swift, D.M. (1987). Pastoralism and drought in Turkana District, Kenya. A report to Norad. Bellevue, CO: Development Systems Consultants, Inc.

Ellis, J.E., Coughenour, M.B. and Swift, D.M. (1993). Climate variability, ecosystem stability, and the implications for range and livestock development. In *Range Ecology at Disequilibrium: New Models of Natural Variability and Pastoral Adaptation in African Savannas*, ed. R.H. Behnke Jr., I. Scoones and C. Kerven, pp. 31–41. London: Overseas Development Institute.

Fleay, M. (1996). Karamoja's District Team and Development Scheme. In *Looking Back at the Ugandan Protectorate: Recollections of District Officers*, ed. D. and M.V. Brown, pp. 20–25. Dalkeith, Western Australia: Douglas Brown.

Fratkin, E. (1991). *Surviving Drought and Development. Ariaal Pastoralists of Northern Kenya*. Boulder, CO: Westview Press.

Galvin, K.A. (1992). Nutritional ecology of pastoralists in dry tropical Africa. *American Journal of Human Biology* 4, 209–221.

Gertzel, C. (1988). *The Politics of Uneven Development*. Discussion Paper no. 20, Flinders University of South Australia, Kampala, Uganda: UNICEF.

GOU (Government of Uganda), Ministry of State for Karamoja Affairs (1994). *All Eyes and National Attention to Karamoja*. Proceedings of a national conference on strategies for peace and sustainable development for Karamoja and neighboring districts. Kampala, Uganda, 18–22 July.

Gray, S.J. (1992). Fertility levels and birth spacing among a group of breastfeeding

128    *S. Gray* et al.

Ngisonyoka Turkana women. *American Journal of Physical Anthropology* **14** (Suppl.), 84.

Gray, S.J. (1998). Butterfat feeding in early infancy in African populations: new hypotheses. *American Journal of Human Biology* **10**, 163–178.

Gray, S.J. and Akol, H.A. (2000). Reproductive histories and life history of Bokora and Matheniko Karimojong women of northeast Uganda. Paper presented at the 69th Annual Meeting of the American Association of Physical Anthropologists, San Antonio, Texas, April 12–15, 2000.

Gulliver, P.H. (1955). *The Family Herds. A Study of Two Pastoral Tribes in East Africa: The Jie and the Turkana.* London: Routledge & Kegan Paul.

Hardin, G. (1968). The tragedy of the commons. *Science* **162**, 1243–1248.

Jelliffe, D.B., Bennett, F.J., Jelliffe, E.F.P. and White, R.H.R. (1964). Ecology of childhood disease in the Karamojong [sic] of Uganda. *Archives of Environmental Health* **9** (July), 25–36.

Johnson, B.R. (1999). Social networks and exchange. In *Turkana Herders of the Dry Savanna. Ecology and Biobehavioural Response of Nomads to an Uncertain Environment*, ed. M.A. Little and P.W Leslie, pp. 89–106. Oxford: Oxford University Press.

Lamphear, J. (1976). *The Traditional History of the Jie of Uganda.* Oxford: Clarendon Press.

Lamphear, J. (1992). *The Scattering Time. Turkana Responses to Colonial Rule.* Oxford; Clarendon Press.

Lamphear, J. (1993). Aspects of becoming Turkana. In *Being Maasai: Ethnicity and Identity in East Africa*, ed. T. Spear and R. Waller, pp. 89–104. London: James Curry.

Lamphear, J. (1998). Brothers in arms: military aspects of East African age-class systems in historical perspective. In *Conflict Age & Power in North East Africa*, ed. E Kurimoto and S. Simonse, pp. 79–97. Oxford: James Currey.

Leslie, P.W. (1996). Turkana population dynamics: demographic response, resilience and persistence in a fluctuating environment. *American Journal of Physical Anthropology* **22** (Suppl.), 148.

Leslie, P.W. and Dyson-Hudson, R. (1999). People and herds. In *Turkana Herders of the Dry Savanna. Ecology and Biobehavioural Response of Nomads to an Uncertain Environment*, ed. M.A. Little and P.W. Leslie, pp. 233–247. Oxford: Oxford University Press.

Leslie, P.W. and Fry, P. (1989). Extreme seasonality of births among nomadic Turkana pastoralists. *American Journal of Physical Anthropology* **79**, 103–115.

Leslie, P.W., Campbell, K.L. and Little, M.A. (1993). Pregnancy loss in nomadic and settled Turkana women in Turkana, Kenya: a prospective study. *Human Biology* **65**, 237–254.

Leslie, P.W., Campbell, K.L., Little, M.A. and Kigondu, C.S. (1996). Evaluation of reproductive function in Turkana women with enzyme-immunoassays of urinary hormones in the field. *Human Biology* **68**, 95–118.

Leslie, P.W., Campbell, K.L., Campbell, B.C., Kigondu, C.S. and Kirumbi, L.W. (1999). Fecundity and fertility. In *Turkana Herders of the Dry Savanna. Ecology and Biobehavioural Response of Nomads to an Uncertain Environment*, ed. M.A. Little and P.W Leslie, pp. 249–278. Oxford: Oxford University Press.

Leslie, P.W., Dyson-Hudson, R. and Fry, P. (1999*a*). Population replacement and persistence. In *Turkana Herders of the Dry Savanna. Ecology and Bi- obehavioural Response of Nomads to an Uncertain Environment*, ed. M.A. Little abd P.W. Leslie, pp. 281–301. Oxford: Oxford University Press.

Leslie, P.W., Little, M.A., Dyson-Hudson, R. and Dyson-Hudson, N. (1999*b*). Synthesis and lessons. In *Turkana Herders of the Dry Savanna. Ecology and Biobehavioural Response of Nomads to an Uncertain Environment*, ed. M.A. Little and P.W. Leslie, pp. 355–373. Oxford: Oxford University Press.

Little, M.A. (2002). Human biology, health, and ecology of nomadic Turkana Pastoralists. In *Human Biology of Pastoral Populations*. ed. W.R. Leonard and M.H. Crawford, pp. 151–182. Cambridge: Cambridge University Press.

Little, M.A. and Leslie, P.W., ed. (1999). *Turkana Herders of the Dry Savanna. Ecology and Biobehavioural Response of Nomads to an Uncertain Environment*. Oxford: Oxford University Press.

Little, M.A., Dyson-Hudson, R., Dyson-Hudson, N. and Winterbauer, N.L. (1999*a*). Environmental variations in the South Turkana Ecosystem. In *Tur- kana Herders of the Dry Savanna. Ecology and Biobehavioural Response of Nomads to an Uncertain Environment*, ed. M.A. Little and P.W. Leslie, pp. 317–330. Oxford: Oxford University Press.

Little, M.A., Dyson-Hudson, R. and McCabe, J.T. (1999*b*). Ecology of South Turkana. In *Turkana Herders of the Dry Savanna. Ecology and Biobehavioural Response of Nomads to an Uncertain Environment*, ed. M.A. Little and P.W. Leslie, pp. 43–65. Oxford: Oxford University Press.

Little, M.A., Gray, S.J., Pike, I.L. and Mugambe, M. (1999*c*). Infant, child, and adolescent growth, and adult physical status. In *Turkana Herders of the Dry Savanna. Ecology and Biobehavioural Response of Nomads to an Uncertain Environment*, ed. M.A. Little and P.W. Leslie, pp. 187–204. Oxford: Oxford University Press.

Majok, A.A. and Schwabe, C.W. (1996). *Development among Africa's Migratory Pastoralists*. Westport, CT: Bergin & Garvey.

McCabe, J.T. (1987). Drought and recovery: livestock dynamics among Ngisonyoka Turkana of Kenya. *Human Ecology* **15**, 371–389.

McCabe, J.T. (1990*a*). Turkana pastoralism: a case against the tragedy of the commons. *Human Ecology* **18**, 81–103.

McCabe, J.T. (1990*b*). Success and failure: The breakdown of traditional drought coping institutions among the Turkana of Kenya. *Journal of Asian and African Studies* **25**, 146–159.

McCabe, J.T. (1994). Mobility and land use among African pastoralists: old conceptual problems and new interpretations. In *African Pastoralist Systems*, ed. E. Fratkin, pp. 69–89. Boulder, CO: Lynne Rienner Publishers.

McCabe, J.T., Dyson-Hudson, R. and Wienpahl, J. (1999). Nomadic movements. In *Turkana Herders of the Dry Savanna. Ecology and Biobehavioural Response of Nomads to an Uncertain Environment*, ed. M.A. Little and P.W. Leslie, pp. 109–121. Oxford: Oxford University Press.

Prins, H.H.T. (1992). The pastoral road to extinction: competition between wildlife and traditional pastoralism in East Africa. *Environment and Conservation* **19**, 117–123.

Reid, R.S. (1992). Livestock-mediated tree regeneration: impacts of pastoralists on woodlands in dry, tropical Africa. Ph.D. dissertation in Range Science, Colorado State University, Fort Collins, CO.

St. Kizito Hospital – Mathany (1994). *Annual Report.* Kampala, Uganda: Moroto Diocese.

Scoones, I. (1993). Why are there so many animals? Cattle population dynamics in the communal areas of Zimbabwe. In *Range Ecology at Disequilibrium: New Models of Natural Variability and Pastoral Adaptation in African Savannas*, ed. R.H. Behnke Jr., I. Scoones and C. Kerven, pp. 62–76. London: Overseas Development Institute.

Scoones, I., ed. (1994). *Living with Uncertainty. New Directions in Pastoral Development in Africa.* London: Intermediate Technology Publications.

Scoones, I. (1994). New directions in pastoral development in Africa. In *Living with Uncertainty. New Directions in Pastoral Development in Africa*, ed. I. Scoones, pp. 1–36. London: Intermediate Technology Publications Ltd.

Scoones, I. (1996). Coping with drought: responses of herders and livestock in contrasting savanna environments in Southern Zimbabwe. In *Case Studies in Human Ecology*, ed. D.G. Bates and S.H. Lees, pp. 175–194. New York: Plenum Press.

Shell-Duncan, B. (1995). Impact of seasonal variation in food availability and disease stress on the health status of nomadic Turkana children: a longitudinal analysis of morbidity, immunity and nutritional status. *American Journal of Human Biology* **7**, 339–355.

Simonse, S. and Kurimoto, E. (1998). Introduction. In *Conflict Age & Power in North East Africa*, ed. E. Kurimoto and S. Simonse, pp. 1–28. Oxford: James Currey.

Smith, Andrew B. (1992). *Pastoralism in Africa: Origins and Development Ecology.* Athens, OH: Ohio University Press.

Spencer, P. (1998). Age systems & modes of predatory expansion. In *Conflict Age & Power in North East Africa*, ed. E. Kurimoto and S. Simonse, pp. 168–185. Oxford: James Currey.

Stenning, D.J. (1959). *Savannah Nomads: A Study of the Wodaabe Pastoral Fulani of Western Bornu Province, Northern Nigeria.* London: Oxford University Press.

Thomas, E.M. (1965). *Warrior Herdsmen.* New York: Knopf.

Vossen, von Rainer (1982). The Eastern Nilotes: Linguistic and Historical Reconstructions. Berlin: D. Reimer.

Wiens, J.A. (1977). On competition and variable environments. *American Scientist* **65**, 590–597.

Wilson, J.G. (1985). Resettlement in Karamoja. In *Crisis in Uganda: The Breakdown of Health Services*, ed. C.P. Dodge and P.D. Wiebe, pp. 163–170. Oxford: Pergamon Press.

# 6 Changing pattern of Tibetan nomadic pastoralism

MELVYN C. GOLDSTEIN AND CYNTHIA M. BEALL

## Introduction

Nomadic pastoralism as a way of life came under increasing pressure during the twentieth century as political, economic and environmental forces marginalized and sedentarized nomadic pastoralist populations in many parts of the world. Until recently, however, this was not the case on the Tibetan Plateau of the Peoples' Republic of China where the inhabitants have continued to maintain full-scale nomadic pastoralism.

The Tibetan Plateau is a vast upland expanse of intersecting mountains and plains that stretches almost 3000 km along its west–east axis from Indian Kashmir in the west to China's western provinces of Qinghai, Gansu and Sichuan in the east and almost 1500 km along its north–south axis. Most of this area is inhabited exclusively by nomadic pastoralists who have resided there for untold centuries, probably for several millennia. Archaeological research in Tibet is still sketchy so it is not possible to say precisely when pastoralism first appeared there. The earliest village site is dated to 7000 BP (Chang, 1986) and historical materials from Tibet's first great kingdom indicate nomads' presence in the seventh to ninth centuries CE.

A salient characteristic of nomadic pastoralism on the Tibetan Plateau is that it continues to function in its original environment. Tibet's nomads have not been pushed into more marginal environments by agriculturalists because the Tibetan plateau's unique high-altitude environment precludes agriculture. At the latitude of the Tibetan Plateau (roughly 30–40N), farming is not possible at altitudes above ~4400 m above sea level (14 500 feet asl) because the growing season is too short and the cold too intense. If pastoralists did not use these upland areas they would remain unused for production. That was true 1500 years ago and remains true today.

Nevertheless, despite the environmental/climatic buffer against alternative uses of the plateau, the past two decades have seen serious questions raised about the ecological state of the Tibetan Plateau and the future of

131

the Tibetan pastoral way of life. The dominant view among Chinese scientists and government officials is that the Tibetan Plateau is being degraded by overstocking that is the result of the irrational herd maximization practices of the indigenous nomads (Chen *et al.*, 1984) and that China should modernize its agriculture and animal husbandry. The claim that a "Tragedy of the Commons" scenario is in progress on the plateau has prompted the development of new herding policies that seek to change core aspects of the traditional adaptation that the government considers irrational. These policies vary somewhat from area to area but generally seek to restrict mobility and transform nomadic pastoralists into something more akin to Western family ranchers. The most extreme example of this approach is the recent program implemented in the northeastern part of the plateau (Qinghai province) where pastures have been privatized and fenced in on a household basis. This is being expanded to the southeast part of the plateau (Sichuan province), and may ultimately be implemented in the Tibet Autonomous Region itself.

Critics of these policies, these authors included, question the assumption of irrationality as well the empirical evidence for overstocking and serious pasture degradation. They assert that traditional Tibetan pastoral practices successfully allowed continuous use of the Tibetan Plateau for millennia and suggest that the new policies themselves are potentially dangerous since they reduce the mobility and flexibility that traditionally allowed nomads to respond effectively to changes in vegetation.

Consequently, despite a lack of competition from agriculturalists, the future of nomadic pastoralism for the over 500 000 pastoral nomads on the Tibetan Plateau is in question. This chapter explores this issue using data from Phala, a community of nomadic pastoralists located in the Tibet Autonomous Region and studied by the authors over the course of 12 years from 1986 through 1998.

### Phala, Ngamring County, Shigatse Prefecture, Tibet Autonomous Region

Phala is a remote pastoral region (a xiang or "township"[1] in the Chinese system of administration) located in the midst of the Tibetan Plateau about 480 km west of Lhasa and 185 km north of the main east–west road. It is surrounded by other nomad groups and one can travel hundreds of

---

[1] Township is the standard translation of xiang, although in rural areas it refers to a series of local farming or herding communities rather than a town in the Western sense.

kilometers to the east or west of Phala and encounter only other nomadic pastoralists. Phala was selected as a research site in 1986 because it was relatively remote and exemplified the traditional pastoral way of life. Access to Phala is difficult since roads are minimal and there are three high mountain passes between it and the main east–west road.

Data were collected with a variety of anthropological methods including participant observation, in-depth interviews, focused surveys on specific topics, systematic direct measurement of milk, meat and fibers and other aspects of production, and local records. The authors lived in nomad camps more than 24 months over a period of years and have observed everyday life at different points throughout the annual cycle.[2]

Phala covers a large, demographically sparse area. Its several hundred square miles contained 57 households and 265 people in 1988. These were divided into 10 or more encampments; the group never comes together into one settlement. The maximum distance between camps is 2–3 days walk on foot, but any given camp is no more than a half-day walk from the adjacent camps.

The Phala nomads herd four kinds of animals: yak, sheep, goats and horses. Yaks accounted for 9% of the herds in 1986, sheep for 50%, and goats for 40%. Horses comprised less than 1% and are used only occasionally for trips. The nomad's yak, sheep and goats provide a wide range of products, some of which are directly consumed (such as yogurt, butter, meat, wool and skins), and some of which are sold or traded for goods the nomads do not themselves produce (such as barley, tea, and cooking pots). Roughly 50% of the nomads' dietary calories derive from grains they secure from farming areas located 15–20 days' walk to the southeast (Goldstein and Beall, 1990).

Like the rest of the Tibetan Plateau, the climate in Phala is extremely harsh. The short growing season lasts only about 3.5 months. Daily temperature lows that are above freezing occur only in July and August. Winter lows range from −28 to −40 °C (−18 to −40 °F). Annual precipitation is about 200 mm (8 inches), roughly 75% of which falls in June, July and August often in the form of sudden snow and hailstorms.

The chief environmental constraint for Tibetan pastoralists is the short growing season. There are no nearby lowland areas to which the nomads can migrate in winter to gain access to fresh grass. Even if the nomads traveled hundreds of miles the growing season would be roughly the same at their destination. Thus, Phala's herd management strategy seeks to ensure that enough vegetation is left standing at the end of the growing

---

[2] For an overview of these nomads see Goldstein and Beall (1990).

season to sustain their herds until the new vegetation appears the following summer. Livestock in Phala must subsist on senescent vegetation for over 8 months of the year and winter forage is the limiting factor in this ecosystem.

The Phala nomads organize pasture use through an annual pattern of two major migrations. Every household (or group of households) has a home-base encampment surrounded by a government-allocated large pasture area where they spend winter, spring and summer (roughly from late December to September). These home-base encampments may contain only one or two households but sometimes contain as many as 10–12 households. Each household has a winter tent pit and corrals at this location. Since the mid-1980s most households have built one-room winter houses of mud bricks.

Throughout the summer months the nomads move their herds daily to different pastures within the home-base area that are within walking distance of their tents. Then, in September, as the growing season comes to an end, all households leave their home-base encampment and make a major move to one of a number of fall camp sites that are generally no more than a day's walk away. These fall sites are normally at the same or higher altitude than the home-base and have been left ungrazed all summer so they provide excellent forage. Phala is, therefore, basically a horizontal rather than a vertical system of herding.

The nomads remain at these fall sites until the vegetation there is exhausted at which time they return to the main home-base encampment. This usually occurs at the end of December. From then until the start of the next growing season in late May/early June, the animals remain in the general home-base area. Fodder is not provided to the livestock in any significant amounts so the animals graze exclusively on senescent vegetation. Once a year the nomads in Phala travel to a haying area that has been set aside and left ungrazed for the year. However, hay yields are low and the hay is used only as a supplement for horses and for weak sheep and goats.

The Phala nomads' basic migration system is actually more complicated than described above because households frequently split off parts of their herds and send them with satellite tents to separate pastures. For example, in spring, pregnant sheep and goats go to a birthing camp that has been left ungrazed all year and herders set up satellite camps for yaks higher in the mountains in winter. At other times during the year satellite camps are used to adapt to local climatic conditions; for example, one or two family members may move with a portion of the herd to a distant pasture to take advantage of ungrazed forage or a different species of vegetation.

This system of migration enables the nomads to use the growing foliage at their home-base area during the summer, use previously ungrazed pasture in a different location during the fall and early winter, and then return to their summer home-base area where a full cover of senescent foliage has been preserved by the fall move that will suffice for the 5 months of winter/spring from late December to late May.

Tibetan nomads place high value on the number of animals they own and all households seek to maximize the size of their herds. While at first glance this seems to support government criticisms of the traditional pastoral management system, it actually appears to be a rational response to a nonequilibrial physical environment.

### Phala as a nonequilibrial but persistent system

Nonequilibrial systems are characterized not by a system of self-regulated homeostasis but rather by great fluctuations in the number of livestock as a result of random, unpredictable and uncontrollable natural calamities that periodically decimate herds (Ellis and Swift, 1988; Little *et al.*, 1999). In nonequilibrial systems, herds in an area typically increase rapidly for some period of time and then decline precipitously as a result of an externally imposed disaster. Thus the number of livestock in an area varies widely through time but there is no long-term pattern of exponential increase in the number of animals. On the Tibetan Plateau, these calamities are mainly caused by unusually heavy winter snows that cover vegetation and prevent grazing, although disease and low rainfall contribute too. For example, during the winter of 1997–98 heavy snowfall in several regions of the Tibetan Plateau was followed by freezing temperatures that prevented the snow from melting and prevented the animals from access to the vegetation. The result was the loss of several million head of livestock in the Tibeten Autonomous Region (TAR). Some areas, for example, lost 50% of their livestock (Miller, 1998). This pattern of recurrent episodes of livestock decimation appears to typify the Tibetan Plateau. It has created a nonequilibrial, persistent system in which the grasslands were not systematically overgrazed despite a strong motivation for maximizing herd size and continuous use for at least one, and perhaps two or more, millennia.

The unpredictability of natural disasters over large expanses of territory is replicated at the local level in the sense that the areas used by individual herders are also subject to random calamities even though the larger geographical area is not severely affected. For example, one household might experience a 95% survival rate of newborns in a given year while an adjacent household might experience only 25% survival. Consequently, the rational choice for an individual herder is to maximize the size of his herd at all times since there is no way for him to predict when a natural disaster will decimate his herd and no way for him to prevent it. Facing the possible loss of a large proportion of one's herd due to unpredictable and unpreventable phenomena, it is clearly more advantageous to possess 500 sheep rather than only 50 or 100. The ultimate danger for a pastoralist is falling below the minimum number of animals needed to support the household after a disaster (or a series of bad years) and to recoup herd size quickly during the more advantageous years. To fall below this level in traditional Tibetan society inevitably meant losing one's autonomy and status as an independent herder and being forced to subsist as a laborer for a wealthy herder. Thus, the traditional Tibetan pastoralist's emphasis on maximizing livestock numbers was actually a rational strategy for minimizing risk in an uncertain, nonequilibrial ecological system.

Because of the recurrent fluctuation in herd size, some nomads inevitably experienced large increases in herd size while others experienced large losses. The Tibetan pastoral system adapted to these inevitable fluctuations by allowing movement of herds from areas where stocking levels have risen to ones where they have decreased. Systems of pasture accommodation in traditional Tibet ranged from grazing commons where all members of a group were free to use any pastures in their area, to more formal systems of pasture allocation and reallocation such as that used traditionally in Phala until 1969.

Prior to 1959, the Phala nomads were part of Tibet's feudal economic and political system. They were subjects of one of Tibet's greatest religious lords, the Panchen Lama, and part of one of his huge pastoral estates comprised of thousands of square miles and 10 large administrative units, the smallest of which was Phala. Nomad households held usufruct rights to certain pastures on a 3-year rotating basis and paid taxes to their lord based on the size of their pasture/herd. Every pasture in this huge area was named and allocated a carrying capacity in terms of animals. This carrying capacity was based on the quality of the vegetation rather than any areal measure. The basic unit of measure was the *marke* (in Tibetan this is literally "butter measure"), each unit of which was rated as capable of supporting 13 yak. In turn, six sheep or seven goats were considered equal to one yak so one marke of pasture area had a carrying capacity of 13 yak, 78 sheep or 91 goats (or some combination of these). Horses, newborn animals and stud animals were not counted in these tallies.

Every 3 years the lord conducted a head count of all animals and reallocated pasture areas on the basis of the changes in household herd size. A household whose herds had decreased since the last count would lose access to an equivalent number of marke of pastures. A household whose herds had increased would be awarded access to additional pasture. The variation in herd size due to random fluctuations usually enabled the lord's officials to allocate additional pastures that were near the current pastures. However, in extreme cases where a whole area had increased in the number of livestock, some families were moved to a more distant area where there had been a decrease in animals. This system, therefore, managed fluctuations in herd size in different areas by periodically adjusting the number of animals to pastures over a large area encompassing thousands of square miles.

This system was operative when Tibet was incorporated into the People's Republic of China in 1951. It continued to function until nomad communes were implemented in 1969, although the feudal lord system had ended in 1959. During the commune system, households were no longer units of production. Instead, each individual nomad worked for the commune and received "work points" that were converted into food and clothing, etc. However, the same herd movement system was used under the commune as in the traditional era. Although several programs to increase vegetation via fencing pastures and irrigation were tried during the commune years, these failed. No attempt was made to settle nomads or import farmers into the area. The basic relationship of livestock to pastureland, therefore, continued although herd mobility was now limited to one's own commune rather than to anywhere within the 10 groups previously owned by the Panchen Lama in the old society. This meant that the Phala nomads' commune had only about 10% as much potential pastureland within which to make adjustments as the lord had previously at his disposal.

Another major transformation occurred in 1981 when nomad communes were disbanded in Phala as part of the new nationwide socialist market economic system implemented by Deng Xiaoping. The new Household Responsibility System divided all commune livestock equally among households on a per capita basis. Each nomad in Phala received on average 39 animals (4.5 yak, 27 sheep, and 7.5 goats) that he owned and could do with as he pleased. The question of pastureland reallocation was more problematic.

The nomads would have preferred to revert to the flexible marke rotation system of the pre-communist era wherein all the pastures in Phala were reallocated every 3 years based on changes in the number of livestock. However, the government chose not to do this. Instead it opted for a simpler but less flexible system that divided up Phala's households and pastures into two units called *trongdzo* (communities) and then subdivided the trongdzo into ten subunits called *dzug*. Dzug comprised 2–12 households who jointly had usufruct right over a specified number of pasture areas. These were the basic pasture units. Consequently, although the nomads individually owned their own animals, pastures were not held in common by the whole group as in the traditional era and there was no mechanism for reallocation since rights to pastureland could not be sold or rented either at the household level or at the level of a dzug. If a particular dzug's livestock totals increased or decreased by 50% their allocation of pasture was not increased or decreased, although local officials sometimes worked out informal sharing agreements. Thus, the amount of pastureland within which households could move to accommodate fluctuations in livestock numbers was in effect reduced again. This time the reduction was to 10% of the area available during the commune period and less than 1% of that available during the old society.

Nevertheless, after the dissolution of the commune, the nomads again were operating with the household as the basic unit of consumption and production and day-to-day management decisions rested with each nomad household. In this sense the post-collective era was a partial return to the traditional system of management and production – without the critical pasture reallocation flexibility.

Although the nomads were initially encouraged to increase their herds and "get rich," that policy did not last long. In June 1987, the prefectural and county level governments of the TAR decreed that nomadic pastoral and agricultural communities must reduce their livestock numbers by December of that year. Phala was ordered to reduce herds by 20%. In other words, 20% of the livestock in Phala had to be killed or sold. This new pastoral system operated in a top–down fashion with stocking limits being passed down from the prefecture to the county to the township and ultimately to the individual household. Livestock limits have been in place ever since then.

The rationale for this action was an assertion that the area had experienced rapid livestock population growth since decollectivization in 1981 and that this was causing overstocking and degradation of the grassland. In the words of the governor of Ngamring county, "After decollectivization, between 1981–83, the number of livestock increased and 40% of the pastureland got degraded." (Interview, 7 June 1997.)

However, the Phala nomads and their local officials disagreed with this assessment and, to the contrary, insisted that overall there was excess grass. As the head of one township told us, "We should not keep more animals than the carrying capacity can accommodate, but here we have excess grass at the end of the year so we should be permitted to keep more animals. . . . The county government has never done a survey of the grass to see if it is enough and I think they should do real research and then decide based on the real carrying capacity." (Interview, 13 June 1997.)

Although local nomad leaders have not been able to persuade the county and prefecture to alter their policy, they have tried to soften the impact of the limits by allowing some households to increase if others in the dzug pasture unit have decreased and the overall pasture unit's totals did not rise. They also have allowed a flexible timetable for counting livestock. Initially the reduction number was given in November or December of the year, but that was difficult for the herders since they were reluctant to slaughter or sell a lot of animals at the start of winter when they might have a bad winter/spring and lose a large portion of their herd. Consequently, local officials have modified the timetable so that the actual counting and setting of limits is done in June when the nomads know clearly how many of their adult and newborn animals have survived. Then the nomads are given until the following November to adjust to that limit, giving them the entire summer to sell or trade or slaughter excess animals. Nevertheless, despite this flexibility, the area's overall growth is limited for reasons the local herders do not understand given their perceived excess of pasture. The question this dispute raises – the state of the grasslands on the unique Tibetan Plateau – is important and we examined the issue in 1987–88 and again in 1997–98 in the Phala area.

### Herd growth, myth or reality?

The need to balance herd size with available forage is fundamental to ecological conservation and sustainable production. A crucial question in Phala's case is whether their herds were actually increasing rapidly as the county government claimed. Table 6.1 shows animal census data (based on official local records and household interviews) beginning with the year of decollectivization.

The characteristic fluctuations in herd size that typify nonequilibrial systems can be seen in the 25% decrease in livestock numbers between 1981 and 1985 and then the 27% increase in livestock between 1985 and 1987. Table 6.1 also shows that for the period 1981–87, Phala experienced an overall 4% decrease, and another 4% decrease the year following the imposition of herd limits.

Consequently, despite the nomads' deep-seated value on livestock maximization, empirically, 6 years after the dissolution of the commune Phala

Table 6.1. *Number of animals in Phala, 1981–88*[a]

| Year | Total head of lifestock | Decrease from 1981 | |
| --- | --- | --- | --- |
| | | Number | Percent |
| 1981 | 10 787 | – | – |
| 1983 | 9534 | −1253 | −12% |
| 1984 | 8473 | −2314 | −22% |
| 1985 | 8124 | −2663 | −25% |
| 1987 | 10 335 | −452 | −4% |
| 1988 | 9934 | −853 | −8% |

[a] These data derived from handwritten records found at the township and household interviews.

contained fewer animals than at the end of the commune. Given these data, a puzzling question was why county and prefecture officials were asserting that there had been a substantial increase in stocking rates over this period. It is difficult, of course, to know for certain, but the likely reason is that county officials had incorrect information on changes in herd size between 1981 and 1987 because their records inadvertently under-enumerated the number of livestock in 1981, the baseline year of the commune dissolution.

At the time of commune dissolution, all the animals owned by the commune were divided equally among the members and recorded in official records. The official records, however, did not count the animals held privately by each household during the commune. In Phala, those private animals totaled about 1800 goats, or about 20% of the commune total. Since these animals remained the private possession of nomad households at the time of decollectivization, the official animal total for 1981 was really about 20% lower than the actual number. However, when an area head count was made in 1982 it included these private animals in the total so that even if there were no real increase in the number of commune livestock, the overall total would have been 20% higher. Thus, while the county records show an increase in herd size of 15% from 1981 to 1987, this is simply a spurious artifact of the base year being undercounted in the records. Thus, it is easy to understand the dismay of the Phala nomads at the 1987 government edict imposing limits on herd growth.

Still, even this stocking level could be too large for the Phala environment and could be causing erosion and degradation. Thus, a basic question in Phala (and throughout the Tibetan Plateau) is whether or not the rangeland is in the process of being seriously degraded. Although it is difficult to assess rangeland status, our data do not support the government's contentions. Several lines of evidence indicate that the number of livestock in Phala does not exceed available forage and that the pasture areas are intact.

First, ungulates such as antelope, wild asses, and gazelles were abundant in the area and were seen daily around the nomad's campsites. Such coexistence is not generally found in areas where severe overgrazing has degraded rangeland because wildlife usually lose in any competition with domesticated livestock for vegetation.

Qualitative indicators of overgrazing were also absent. Plant communities were rich in species. In 1987 a rangeland specialist (R. Cincotta) joined our field team to investigate the pastureland and found no evidence that serious degradation was occurring. He collected over 75 species of herbaceous plants from actively grazed rangeland and speculated that more are to be catalogued in other grazing areas. Such variety is generally not found in seriously overgrazed areas. In most Phala pastures nearly every perennial grass plant could attain seed-bearing stage in 1987, which also suggests that degradation of the vegetational component was probably not occurring. Nor was severe erosion or soil compaction visually evident. Although large areas of Phala are sparsely vegetated, plant density appeared to be a function of seasonal soil moisture and soil texture rather than grazing intensity.

Furthermore, we were able to establish baseline information from grazing exclosures. Significant removal of vegetation after one summer of grazing could be detected in only one of the four major vegetation types sampled, and remnant vegetation in this type was 50% of the ungrazed standing biomass, an amount that equals an often recommended limit for grazing. However, the livestock pass through this area twice daily on their way to water so it would be heavily grazed and trampled even under the most careful grazing management (Cincotta *et al.*, 1991).

We also found that the nomads in Phala had excess pasture areas, i.e., pastures that they chose not to use at all in 1987–88 because they were further from their homebase and they decided that the vegetation in the nearby areas was adequate.

Finally, one could ask, if one were to agree with the flawed government statistics, what new factors might account for such a sudden explosion of herd size, given that Tibetan nomad culture has always emphasized maximization of herd size. Climatically there was general agreement in Phala that the years following decollectivization were drier than usual, i.e., that the vegetation was below average. However, that would have led to smaller herds, if anything. Veterinary services in Phala were rudimentary and unlikely to have accounted for any great explosion of herd size. Thus no new influences permitting larger than usual herd growth seem to have operated.

Consequently, our research found no evidence to support the government's claim of overstocking and environmental degradation and therefore no scientific justification for the imposition of limits on herd size. In fact, rather than overstocking, we found the spatially restrictive and inflexible pasture system that was established after decollectivization a major problem. It was creating potentially serious imbalances that posed a danger for the sustainability of the grasslands because some herding units (dzug) were experiencing increases in herd size and were complaining of too little pastureland for their livestock while others had excess forage but no official mechanism for adjustments. Sometimes local leaders could broker arrangements to share pastures for a limited period of time, but the underlying structure of the new system was flawed. In Phala, the imposition of a policy that established limits on herd size controlled this problem, but only at the expense of economic improvement. If such limits were genuinely necessary to preserve the Tibetan Plateau as a rangeland resource for the future such policies would be justified, but looking at the evidence from Phala, it seemed arbitrary and irrational.

### Revisiting Phala in 1997–98

In 1997, 11 years after our first visit in 1986, a follow-up stint of fieldwork in Phala was conducted by the original research team (CMB and MCG) together with a different rangeland scientist (D. Miller). Much had stayed the same, but there were a number of significant changes.

Administratively, Phala had been dissolved as part of a China-wide government program of creating larger and more effective township units. Phala's people and pastures were divided between two adjacent larger townships (Kunglung and Nyingo): 3 of the 10 previous pasture subunits (dzug) were joined into one village (tsön) and placed under Nyingo township, and the rest were organized into a tsön under Khunglung township. Within each of these tsön, the units that had been called dzug now were renamed as *dzasho*.

This has had a marked impact on social interaction, the halves of what had been Phala now have very little interaction. However, in and of itself, it had little impact on the pastoral system *per se* since the pasture-sharing units remained basically the same. Daily and seasonal herding patterns also did not change in any significant way. With regard to the number of

Table 6.2. *Average number of livestock per household and number of animals per capita for Shagar in Phala*

| Year | Measure | Mean | Median | Minimum | Maximum |
|------|---------|------|--------|---------|---------|
| 81 | Total no. of animals/household | 265.44 | 261.00 | 40.00 | 501.00 |
|    | Animals per capita[a] | 41.56 | 40.82 | 38.83 | 50.86 |
| 86 | Total no. of animals/household | 289.11 | 212.50 | 60.00 | 782.00 |
|    | Animals per capita | 49.11 | 42.86 | 23.82 | 97.14 |
| 87 | Total no. of animals/houshold | 275.25 | 191.50 | 60.00 | 842.00 |
|    | Animals per capita | 47.04 | 43.90 | 20.50 | 90.57 |
| 88 | Total no. of animals/household | 208.86 | 137.50 | 43.00 | 689.00 |
|    | Animals per capita | 37.44 | 35.23 | 15.00 | 78.29 |
| 97 | Total no. of animals/household | 173.39 | 131.00 | 32.00 | 647.00 |
|    | Animals per capita | 32.03 | 28.35 | 11.86 | 107.83 |

[a] Including animals privately owned.

livestock and the state of the rangeland, the follow-up data supported our previous findings. Once again we found no evidence of extensive or worsening rangeland degradation. The rangeland specialist who examined vegetative cover and grazing intensities phrased his report as follows: "Rangelands were not overgrazed" and "in 1997 were in good–excellent shape and could have supported even more livestock."[3] For example, in assessing standing forage from the previous year's growth in late spring/ early summer, he found that "in the *Stipa* rangelands of Phala there were still considerable standing crop available from last year's growth, even in close proximity to the nomad settlements. Fifty percent of all *Stipa* plants still had a seed stalk, and many of the seed stalks that had been removed were probably due to wind during the winter and not because of livestock grazing." Again, both nomads and their leaders said that the area had excess pastureland.

Data on livestock numbers for the portion of the Phala that is now a part of Nyingo tsön, moreover, revealed a substantial decrease since 1987–88, as can be seen in Tables 6.2 and 6.3.

Thus, in the decade since limits were imposed and almost two decades after decollectivization, there was no evidence of rapid livestock increase and rangeland degradation. To the contrary, the stocking levels were lower than two decades earlier at the time of decollectivization, and lower than in 1987 and 1988. Of the households in this area, 81% had fewer animals per capita in 1997 than in 1987.

[3] Daniel Miller, unpublished report on 1997 field study in Phala.

Table 6.3. *Total number of animals for Shagar in Phala*

| Year | Total no. of animals | % changes in total no. of animals from 1981 |
|------|------|------|
| 1981 | 4778 | – |
| 1986 | 5204 | 8.92 |
| 1987 | 5505 | 5.78 |
| 1988 | 4595 | − 16.53 |
| 1997 | 3667 | − 20.20 |

Nevertheless, important changes were under way as the TAR government was moving to change the traditional system of herding and pastures in keeping with national priorities and guidelines on animal husbandry.

### Modernizing pastoralism

In general, there is a broad consensus in China that the ultimate goal for Tibet's nomadic pastoralists is a program that will privatize pastures on a household basis (or in some areas a group of several families), fence in winter pastures and haymaking areas, construct winter barns and houses, and use corrals to grow forage crops such as oats.

The privatization program is modeled on modern Western animal husbandry and assumes that by giving individual herders their own pasture resources the herders will individually take steps to protect and improve their property. The program basically seeks to convert open-rangeland herders into subsistence ranchers who will invest in maintaining and improving their pastures and maintaining high-quality (but smaller) herds from which they will produce a portion of their output for subsistence (as in the traditional era), and the rest for sale to government and private markets.

This policy has been most completely implemented in the northeastern section of the Tibetan Plateau in Qinghai Province where implementation began in the mid-1980s. Beginning with winter pastures, each household was given land it could use for the next 50 years. This was not real privatization since the herders did not own that land and could not sell it to other herders. They could only lease their excess pastureland to other households (within the same township) for a period of 1 year.

To facilitate this transformation, the Qinghai government developed a package of new inputs for the herders and is investing large amounts of capital (mainly provided by the central government) to make these available to them. A three-part pastoral development program accompanied the privatization policy that sets out for herders the innovations recommended to utilize optimally their new "privatized" resources. These innovations are, in addition to permanent houses:

1. The fencing of roughly 500 mu (1 mu = 0.1647 hectares) of the best winter pastureland to provide forage during the lean months of spring (or when livestock appear weak and in need of more food during the senescent period).
2. The construction of animal barns/ shelters to be used during the harshest times of winter and spring.
3. The sowing of 5–10 mu of fodder in traditional corrals during the summer when these are not used by livestock.

Although this sounds like a modern approach, it is really an awkward top–down remnant of socialist economic thinking in that unlike modern systems of animal husbandry in the West, there is no mechanism for adjustment. Pastures cannot be bought or sold and families with three or four children (as is common in Tibetan areas) are forced to divide their land between their children, setting up conditions for overgrazing or impoverishment as these new families themselves have multiple children.[4] It is also extremely expensive to implement.

A striking feature of the Qinghai program is its enormous cost. At the heart of the modernization program is providing herders with the means to protect and improve their pastures by controlling access to their land, and this requires fences. In 1996, fencing and a barn cost about 35 000 yuan (roughly $4000 US) per household[5]. Such capital expenditure was well beyond the economic capability of rural herders in even the richest nomad counties in Qinghai so the government has provided funding to assist herders to acquire these inputs directly as subsidies and indirectly via very low interest loans. Table 6.4 reveals the expenditures for fences in one poor

---

[4] Minority areas like Tibet are permitted much larger families than Han Chinese, so it is common for women to have 4 or 5 children in Tibet and Qhinghai. For a discussion of fertility in Tibet see Goldstein and Beall (1991*a*).
[5] 8.2 yuan = 1 US dollar. To put this in perspective, the per capita income in the richest of Goulou Prefecture's 51 townships was only 2736 yuan.

146     M.C. Goldstein and C.M. Beall

Table 6.4. *Expenditures for fences in Dari County, 1990–95*[a]

| Year | No. fence sites | No. mu fenced | Total investment (yuan) | Cost of fences (yuan/mu) | Gov. investment (yuan [%]) | Private investment (yuan) |
|---|---|---|---|---|---|---|
| 1990 | 35 | 20 000 | 168 000 | 8.4 | 68 725 (41%) | 99 275 |
| 1991 | 8 | 10 000 | 70 000 | 7 | 45 396 (65) | 24 604 |
| 1992 | 70 | 35 000 | 245 000 | 7 | 210 700 (86) | 34 300 |
| 1993 | 80 | 40 000 | 1 704 640 | 42.6 | 1 278 430 (75) | 426 160 |
| 1994 | 90 | 33 000 | 1 225 350 | 37.1 | 1 041 548 (85) | 183 802 |
| 1995 | 114 | 32 500 | 1 396 500 | 43 | 1 187 025 (85) | 209 475 |
| Total | 397 | 170 000 | 4 809 490 | 24.2 | 3 831 824 (80) | 977 616 |

[a] From Goldstein (1998).

county (Dari) in Goulou Prefecture (in Qinghai), where the program started in 1990.

Table 6.4 reveals that the government has invested almost 4 million yuan ($467 000 US) in fencing since 1990 as direct subsidy and that this accounted for 80% of the total cost. Another 1 million yuan was spent in subsidizing winter barns, bringing the total in just one county to over $600 000 US, and the program was far from complete. Moreover, the nomads themselves had invested 1.4 million yuan ($170 000) in fences and barns as of 1995 via low interest loans. However, it is not at all clear whether the poorest families will be able to afford even the low cost loans for these, and whether the government will be able to subsidize completely their fences and barns. If not, then the specter of some unfenced pastureland being destroyed by overuse by herders other than those who have rights to it is a serious danger (Goldstein, 1998). This has apparently occurred in parts of Inner Mongolia where a similar pasture system was implemented (Williams, 1996).

The Qinghai pasture privatization program was recently begun in Sichuan province (Wu and Richard, 1999) and was considered in Phala and the TAR. However, it was rejected – temporarily – as too expensive. This level of expenditure was not feasible in the Tibet Autonomous Region where individual households could not afford subsidized loans and where most were far from market where they could easily sell their goods. For example, in Phala in 1997, 50% of the households had fewer than the 30 animals per capita level that the government used to demarcate poverty. Consequently, a less ambitious program of pasture reallocation and poverty alleviation was implemented in its place.

In Phala, the first step occurred in 1991–93 when all of the pasture areas were finalized with regard to their carrying capacity. Since the county was unable to determine size and the carrying capacities in modern units (e.g., mu or /hectares), it decided to use the marke tax units from the feudal period. This provided a relative carrying capacity for each pasture area, and allowed each community to establish the total number of marke units in their area.

The new stocking limit (the number of animals permitted per marke) of each tsön was artificially calculated by dividing the number of animals present in 1966 – the year set as the index year – by the number of marke units. Thus, if a community happened to have fewer animals in the 1996 index year because of a bad winter, its per marke carrying capacity was lower than neighboring areas and this amount became its permanent stocking limit.

Despite the local government's use of marke, the flexible reallocation dimension of the traditional system was not reinstated. Instead, the pastureland was redivided within each tsön based on the number of animals and people present in each pasture unit in 1996 (50% based on animals and 50% based on people in some townships and 60% on people and 40% animals in others). The notion underlying this was not so much optimizing animal husbandry but rather socialist social engineering – to give poor nomads with large families and few animals a greater share of pastureland. Thus, if a pasture unit had 100 marke, 60 marke was divided on the basis of the number of people in the pasture unit in 1996 and 40% on the number of animals.

As in the past, each pasture unit could not exceed its limit of animals and within a pasture unit each household had a limit based on the size of its herd in 1996. However, the new policy was significantly different from the previous one in the sense that in the past households could move freely within their pasture group's area and were permitted to increase herd size so long as other households decreased to an equal amount and the total unit's numbers remained stable. Under the new policy, alleviating poverty was added as an important goal and households within a pasture unit whose herds had increased had to pay a pasture fee (tsarin) to households in that pasture unit whose herds had decreased because the former were "using" the latter's share of the pastureland. Similarly, herding groups were said to be free to lease pastureland to other herding groups for a fee, although this was still in the talking stage as of 1998. Similarly, the pasture fee for each excess sheep had not been settled in 1998. Initial figures of one yuan per animal were rejected by the richer herders who quietly threatened to kill or sell all their excess animals rather than pay such a high fee to the poor whom they characterized generally as lazy and incompetent.

Along with these changes the government also implemented programs to test plant fodder and food (turnips, barley) and to begin fencing some key winter/spring pastures, although as of 1998 these were still mainly token. These changes, although much less radical than those begun in Qinghai, appear to be headed in the same direction. The main impediment to wide-scale fencing in Phala and other TAR nomad areas appears to be funding rather than government agreement with the herders' contention that open pastures and reallocation systems are the most effective management system for the Tibetan Plateau.

## Conclusion

Nomadic pastoralism on the Tibetan Plateau is the only viable subsistence technology to exploit the vast areas above 4,400 m (14,500'). Thus, while Tibetan nomads such as those in Phala are not threatened by the encroachment of farmers, their way of life is being transformed as a result of government policies that are aimed at modernizing the traditional pastoral management system and by social engineering. However, it is not at all clear that these policies are scientifically valid and an improvement over traditional practices. In fact, our empirical field research has revealed no evidence of widespread environmental degradation nor any evidence of an explosion of stocking rates in the areas we studied. The rangeland in Phala was intact in 1987–88 and 1997, and nomads and local officials concurred that there was excess vegetation both years.

In fact, many of the "innovations" of the governments new program were already part of the traditional herd management system. For example, the goal of fencing is to set aside certain areas during the growing season for use in winter/spring. Tibetan pastoralists in Phala traditionally used this principle with respect to their spring birthing and fall pasture areas which they left ungrazed until they were used. There were no fences, but the idea that it was beneficial to set aside an area for use at the time of lambing or for their fall move was part of their traditional system and was apparently successful. Similarly, the notion of sheltering animals in the coldest parts of winter was used. In winter, sheep and lambs traditionally were crowded into walled (albeit uncovered) corrals at night to reduce cold stress. Moreover, in Phala, for a few weeks after birth, lambs were placed at night into small boxes constructed of dung or stone or sod bricks that were covered with a skin or woven cloth to provide a warmer environment than the corrals. Similarly, Tibetan nomad groups understood the efficacy of storing fodder for winter/spring and set aside natural grassland areas appropriate for hay-cutting that were harvested in fall when the grasses were at their maximum height.

It is, therefore, not at all clear that the government's new program for rationalizing and modernizing Tibetan nomadic pastoralism is an improvement over the traditional system. It is also unclear whether the new changes will inadvertently precipitate degradation of the rangeland because the system dramatically restricts the inherent mobility that typified the traditional system and provides no mechanisms for adjusting pasture land to animal population growth. The future of Tibetan nomadism, therefore, is uncertain. There will always be animal husbandry on the Tibetan Plateau simply due to the lack of alternatives. Whether this will be practiced by nomads or by sedentarized, family ranchers is not at all clear today. What is sorely needed is a major scientific program to assess the state of the Tibetan Plateau's rangelands and develop sound models of human–environment–animal interactions there before some "careless technology" destroys this unique ecosystem.

### References

Chang, K.-C. (1986). *The Archaeology of Ancient China, 4th edn.* New Haven and London: Yale University Press.

Chen, H., Shi, N. and Shi, S. (1984). *Agricultural Geography of Xizang (Tibet).* Beijing: Science Press (in Chinese).

Cincotta, R.P., Soest, P.M.V., et al. (1991). Foraging ecology of livestock on the Tibetan Changtang: a comparison of three adjacent grazing areas. *Arctic and Alpine Research* **23**, 149–161.

Ellis, J.E. and Swift, D.M. (1988). Stability of African pastoral ecosystems: alternative paradigms and implications for development. *Journal of Range Management* **41**, 450–458.

Goldstein, M.C. (1998). Report on Qinghai Livestock Development project (manuscript).

Goldstein, M.C. and Beall, C.M. (1989). The Impact of China's cultural and economic reform policy on nomadic pastoralism in western Tibet. *Asian Survey* **29**, 619–641.

Goldstein, M.C. and Beall, C.M. (1990). *Nomads of Western Tibet: The Survival of a Way of Life.* Berkeley, CA: University of California Press.

Goldstein, M.C., Beall, C.M. and Cincotta, R.P. (1990). Traditional nomadic pastoralism and ecological conservation on Tibet's Northern Plateau. *National Geographic Research* **6**, 139–156.

Goldstein, M.C. and Beall, C.M. (1991*a*). China's birth control policy in the Tibet Autonomous Region: myths and realities. *Asian Survey* **31**, 285–303.

Goldstein, M.C. and Beall, C.M. (1991b). Nomadic pastoralism on the Western Tibetan Plateau. *Nomadic Peoples* **28**, 105–122.
Little, M.A., Dyson-Hudson, R., *et al.* (1999). Ecology of South Turkana. In *Turkana Herders of the Dry Savanna: Ecology and Biobehavioural Response of Nomads to an Uncertain Environment,* ed. M.A. Little and P.W. Leslie, pp. 43–66. Oxford: Oxford University Press.
Miller, D. (1998). Hard times on the Plateau. *China Development Brief*: 1(1).
Miller, D. (1999). Nomads of the Tibetan Plateau Rangelands in Western China, Part 3: Pastoral development and future challenges. *Rangelands* **21**, 17–20.
Williams, D.M. (1996). Grassland enclosures: calalyst of land degradation in Inner Mongolia. *Human Organization* **55**, 307–313.
Wu, N. and Richard, C. (1999). The privatization process of rangeland and its impacts on the pastoral dynamics in the Hindu-Kush Himalaya: the case of western Sichuan, China. In *People and Rangelands Building the Future*, VIth International Rangeland Congress Proceedings, Vol. 1, pp. 14–21, Townsville, Queensland, Australia.

# 7 Human biology, health, and ecology of nomadic Turkana pastoralists

MICHAEL A. LITTLE

## Introduction

East African pastoralists of the Sudan, Ethiopia, Kenya, Uganda, and Tanzania constitute a widespread and ethnically diverse group of populations. Some populations are sedentary and practice mixed cultivation and herding, while others are semi-nomadic or nomadic and rely almost exclusively on the products of their domestic livestock for subsistence. As noted by Gray et al. (2002), the Turkana of northwest Kenya are one of the last remaining populations to practice a form of nomadic pastoralism in East Africa. Most other nomadic or semi-nomadic populations have been forced to settle with their livestock or to participate in sedentary group ranching because of the constraints of population growth and modernization in these savanna lands.

In the late 1970s, the South Turkana Ecosystem Project was initiated to build on earlier ethnographic work by Gulliver (1955) and Dyson-Hudson (1973), and to apply multidisciplinary and longitudinal research approaches to studies of nomadic Turkana. The multidisciplinary work was conducted for a decade throughout the 1980s, whereas more specialized topical research continues to the present. The project investigations combined the efforts of rangelands ecologists, sociocultural anthropologists, biological anthropologists, and biomedical scientists to study the Turkana. Research was based on theoretical frameworks that encompassed evolutionary and ecological concepts of adaptation to the environment, and ecosystems approaches to understanding the behavior and biology of the people. Within this framework, the human and livestock populations are viewed as integrated components of the savanna ecosystem. A detailed report of this research can be found in Little and Leslie (1999).

The Turkana are a major ethnic population numbering more than 200000, with several ethnic subdivisions or sections, that reside in the semi-arid savanna of northwest Kenya (see Figure 7.1). Some members of most of these ethnic sections of the Turkana practice pastoral subsistence,

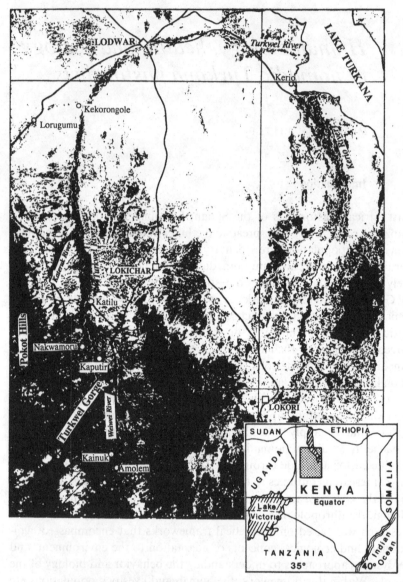

Figure 7.1. A map based on a landsat image of the southern part of Turkana
District (South Turkana) with an insert map of Kenya. The dark areas are either
highlands and mountains or river systems with moderate to substantial
vegetation. Light areas have limited vegetation, except during the wet seasons.
The gallery forest of the Turkwel River, which flows south to north, becomes less
dense as the river passes the District capital, Lodwar, and empties into lake
Turkana. During the dry season, the Turkwel River may cease to flow beyond
the Turkwel Gorge or the town of Kaputir.

by herding camels, cattle, sheep, goats, and donkeys. Residence of the Turkana in northwest Kenya dates back to the 1700s when the Turkana first moved into the region and began a period of expansion and conquest (Lamphear 1988; Dyson-Hudson 1999). Our project centered on the Ngisonyoka Turkana section, a population numbering about 10 000, whose primary subsistence is nomadic pastoralism, and whose territory of about 10 000 km$^2$ is situated in the region of South Turkana roughly circumscribed by the Kerio and the Turkwel rivers (Figure 7.1).

The climate of the Turkana region is driven by seasonal monsoon rainfall that is of low predictability, and where multi-seasonal drought occurs one or more times each decade. Periodic seasonal drought, with annual wet and dry seasons, produces a regular cycle of vegetation flush and senescence. The Turkana nomadic movement of people and herds and their adjustments to food availability reflect distinct adaptations to seasonal drought. Multi-seasonal drought, however, resulting from failure of seasonal rains in one or more years, contributes to a substantial die-off of vegetation and consequent loss of livestock condition or often actual starvation of livestock. These conditions profoundly affect the pastoralists' food supply which is derived largely from the livestock products: milk, blood, and meat. Rainfall and relative availability of green forage for livestock is very difficult to predict. With nearly 20 years of research conducted in South Turkana, it is quite clear that long-term studies were necessary to demonstrate the remarkable degree of year-to-year variation in climate and consequent vegetation availability, and its influence on human behavior, biology, and health.

In addition to the temporal variation in this region, there is considerable spatial variation in patchiness of rainfall and vegetation (Dyson-Hudson 1989). This results from: (1) the scattered, convective monsoon rains, (2) several gradients in rainfall through space (wetter to the west and to the south) and elevation (wetter at higher elevations), (3) subsurface hydrology and stream drainage patterns, and (4) topography, mountains, land surface, and soil variations.

Some of the vegetation in the South Turkana ecosystem is relatively persistent while other forms of vegetation appear and disappear seasonally. For example, the leafy vegetation of acacia trees and tall shrubs tends to persist throughout the year because the deep roots tap below-ground water sources, contributing substantially to the diets of browsing camels (see Figure 7.2). Acacia seed pods, which appear annually, provide an important food for goats, as do dwarf shrubs, which are somewhat drought resistant. Herbaceous plants (grasses and forbs), which are the most ephemeral of kinds of vegetation, appear during the rains, are an

Figure 7.2. A wooded area of acacia trees adjacent to a dry stream bed. The trees provide shade, acacia seed pods for goats, leafy vegetation for camels, thorny branches used for constructing corrals, and materials for constructing huts in the settlement. (Photograph by P.W. Leslie.)

important source of high-protein food for grazing cattle and sheep and then die back after the rains to become unpalatable for all livestock. When it appears, and which kind of vegetation is appropriate for which livestock species, is a part of the indigenous knowledge of the Turkana herders (Coppock *et al.*, 1986*a*; Coughenour *et al.*, 1990*a,b*).

### The Turkana people and their subsistence

Because of the capriciousness of the environment and the severely limited resources, the Turkana people have developed the ability to respond opportunistically when vegetation becomes available for their livestock (Dyson-Hudson, 1980). Their principal opportunistic device is mobility. Their primary means of achieving this mobility is by having a portable material culture, and a sociocultural system that is geared to nomadism. And their ability to make mobility pay off is by their having a detailed knowledge of both their environment and the animals they move through this environment. Other features of Ngisonyoka Turkana culture that are

Figure 7.3. A Turkana man, woman, and child resting in a day hut. (Photograph by M.A. Little.)

adaptive in a semi-arid, highly pulsed, and unpredictable environment are: an intense attachment to their livestock, a drive to maximize livestock holdings, and a complex social system that promotes reciprocity, exchange, and cooperation. One form of exchange is brideprice. Turkana are polygynous and brideprice in livestock must be paid, often amounting to substantial numbers of animals. Other forms of exchange include loans of labor for herding or livestock for food to redress imbalances within families.

The Ngisonyoka Turkana are organized into family herding units that consist of an adult male herd owner, his wife or wives, their children, and, often, other kin (see Figure 7.3). The family settlement or camp consists of thorn bush-enclosed corrals, day huts for working in shade, and night huts for sleeping (see Figure 7.4). Larger social units may be formed of two or more families that move together for purposes of security or sociality. Herd owners who move together may be close kin or simply good friends. Even larger neighborhoods of many camps may congregate when resources are abundant during the rainy season.

Family units move their camps up to 12 times each year in search of forage for their livestock. When moves take place, donkeys are used for transport of vessels, hides, and other portable items of material culture (see

156     *M.A. Little*

Figure 7.4. A night hut (*akai*) in the foreground and a day hut in the background (*ekol*). The night hut will be covered with hides for sleeping during the night; the day hut with its large opening is a comfortable place for women to work during the day. In a polygynous family, each wife will have her own *akai* and *ekol*. (Photograph by M.A. Little.)

Figure 7.5). Herds may be divided into species-specific groups that follow different grazing orbits throughout the year, and these divisions may increase during drought when palatable forage is scarce (Dyson-Hudson, 1989). Grazing orbits among many Ngisonyoka family units are quite regular on a seasonal basis during years with adequate rainfall. During drought years, the regular patterns break down, herds are divided into small units, and grazing orbits both expand and become irregular (McCabe, 1994). When prolonged or multi-year drought strikes the Ngisonyoka lands, the people respond in a variety of ways that include: (1) increased dependence on relatives and friends for support, (2) subsistence shifts and use of unusual foods, and (2) out-migration from the pastoral sector or even the region.

Food produced by livestock includes milk, blood, and meat. Animals are milked daily; ordinarily, only fallow female animals or male adult animals are bled; large stock (camels and cattle) are slaughtered only rarely, whereas small stock (sheep and goats) are slaughtered for meat from time to time. Milk is the staple food, with blood, meat, maizemeal,

Figure 7.5. A Turkana woman and children loading a donkey in preparation for a move. In addition to items of material culture, the woven "saddle" may also be used to transport an infant or young child. (Photograph by J.T. McCabe.)

sugar, and hunted and gathered foods being secondary. Milk production of the livestock is a function of many variables including: (1) forage availability, (2) health and condition of the animals, (3) livestock reproductive state, and (4) season of the year. Milk production of the herds tends to drop during the dry season, and may cease altogether during severe drought. Balances must be struck between the needs for milk of the suckling animals and the nutritional needs of the people. Under most conditions, camels show the least seasonal declines in milk production because of their dependence on arboreal leafy vegetation which is often green throughout the dry season and during drought. Hence, the success of

Figure 7.6. A camel being bled in a corral. The woman is holding the camel tightly by the lower lip which tends to calm the animal. (Photograph by J.T. McCabe.)

a herd owner in managing resources during seasonal or multi-seasonal drought is partly a function of the proportions of different livestock species that are kept. Different livestock mixes require different strategies of management, especially during drought.

The practice of bleeding animals, which is unique to eastern Africa, is done at intervals of 1 to 3 months in animals that are neither pregnant nor lactating. Large camels may produce up to 3 liters of blood at each bleeding episode (see Figure 7.6), while small sheep or goats may produce only 300 ml of blood. Meat consumption, ordinarily relatively low, increases dramatically during drought, when animals begin to starve and die. For example, following the prolonged *Lochuu* drought (1979–81), meat consumption was very high because herds were being depleted by high mortality. When the drought finally ended (1981), milk production was very low because animals were in poor condition and had not reproduced. Later in 1981 and in 1982, most remaining livestock reproduced, which led to abundant milk resources in 1982. Hence, drought effects on livestock condition, reproduction, and milk production may be felt for some time after recovery and restoration of vegetation.

## Turkana health and adaptability

A variety of studies were conducted to assess Ngisonyoka health status. Principal means of assessment included measures of: morbidity and mortality; reproductive health; growth of infants, children, and adolescents; nutritional intake and status of all gender and age categories; body composition; health and immune function; and work capacity and physical fitness. Comparisons of Turkana were made with (1) data on Western populations, (2) other African populations, and (3) samples of Turkana who had given up the nomadic way of life and become settled to take up farming or some other pattern of subsistence. A review of some of this research follows.

### Diet and nutrition

The general picture of nutritional status of the nomads is quite complex, but can be presented in its broad perspective based on several field studies of adults and children (Galvin, 1985; Gray, 1992; Shell-Duncan, 1994). Milk is a staple and highly desired food; blood and meat are supplementary or emergency foods; maizemeal, sorghum, and sugar are also supplementary foods acquired through trade or purchase; some foods are hunted and gathered. An additional source of food is through government and private agency contributions as a part of famine or drought assistance. The proportions of these foods in the diet vary considerably according to season and year, and are especially variable during drought conditions. Although there are generalized dietary patterns, no two seasons nor two years are precisely the same. As noted, during drought when animals are dying, meat (and protein) intake may increase dramatically, but the high cost is in future dietary deprivation because of the depletion of family herds. The mix of livestock species is an important factor in the current and future production of milk for a family (Coughenour *et al.*, 1985). For example, camels are expensive animals but are good milk producers throughout the year; goats and sheep reproduce at high rates such that herd holdings increase rapidly and these small stock can be slaughtered for meat. On the other hand, small stock are time-consuming to milk because milk production is low per animal. Zebu cows are good producers of milk but only during the wet months of the year when grass productivity is high. Hence, dietary adequacy of a nomadic family is a function of the livestock holdings of a herd owner in proportion to the number of his human dependents, his management skills, and the highly variable conditions of vegetation in the environment.

Since many East Africa pastoral and agropastoral populations herd more than one species of livestock, attempts have been made to establish a standard unit for livestock production. One such unit is the Tropical Livestock Unit (TLU), which can be equated to numbers of individuals in a household. The FAO (1967) standards established a camel as equalling 1.25 TLU, a cow, 1.0 TLU, and a sheep or goat, 0.125 TLU. Values of TLU/person for East African pastoral groups range between 3.0 TLU/person and 16.0 TLU/person (Galvin, 1992). The Ngisonyoka Turkana nomads with whom we worked had values around 3.5 TLU/person (Galvin et al., 1994). It must be emphasized that TLU/person values must be viewed as highly variable based on numerous factors (Dyson-Hudson and Dyson-Hudson, 1980). Some of these include: (1) the productivity of the environment and its seasonality, (2) the productivity of the herds, (3) the herd demography (age–sex distribution of animals – females produce milk, males do not), (4) the human demography (age–sex–activity distribution of the members of the human population), and (5) intangibles, such as the fact that a camel is much more highly valued than a cow in a very arid (xeric) environment, whereas a cow is highly productive in a wetter (mesic) environment.

The relatively low average tropical livestock unit value of 3.5 TLU/person for the Ngisonyoka Turkana suggests limited food availability, particularly food energy, which is confirmed by a number of studies. Modeling of dietary energy intake (Little et al., 1988) was conducted based on a model developed by Leslie et al. (1984). The model predicted a dry season intake for all members of the population (mean intake for all age and sex categories) of 1500 kcal/person per day (6280 kJ/person per day) and a wet season intake of 1580 kcal/person per day (6615 kJ/person per day). These values are slightly less than FAO/WHO (1985) energy intake standards, but should be viewed in the context of Ngisonyoka Turkana: (1) age and sex distribution of the population (Little and Leslie, 1990), (2) activity patterns of adults and children (Galvin, 1985), (3) the low body mass of adults and children (Little et al., 1983), and (4) the low resting metabolic rates of adults (Curran-Everett, 1990). Galvin (1985; Coughenour et al., 1985), based on dietary survey of several nomadic Ngisonyoka families, found a mean energy intake of 1375 kcal/person per day (5757 kJ/person per day) that was slightly lower than the values predicted from the model. There is substantial variation in dietary intake by season and by year; hence, these low caloric values should be viewed simply as estimated averages.

Despite the limited energy in the Turkana diet, protein intakes of nomads are substantial by world standards (FAO/WHO, 1985). Galvin (1985) found that protein intake was two to four times daily requirements, with children in the higher range of protein intakes, but all age and sex groups were above 100% of minimum daily needs. Adults will often do without food in order that children might be well fed, particularly with milk, which can explain the high protein intakes in children. Cowives will share food, and women appear to have a slight dietary advantage because they have responsibility for milking the animals (Gray, 1994*a*). Moreover, there is a complex social system of asking for and giving food that entails obligation and trust. These factors all contribute to patterns of distribution of food and variation in dietary intakes and nutritional status.

Seasonal and year-to-year variations in consumption of milk and other foods are substantial. Figure 7.7 illustrates seasonal variations in dietary intakes in late 1981 and throughout most of 1982 (Galvin, 1985). Milk as a proportion of calories in the diet was very high, at greater than 90%, during the wet season. During this year, milk contributed an average of more than 60% of food calories to the diet. As noted earlier, however, this year might have had extraordinary milk production because animals were recovering from the *Lochuu* drought and had had a previous year of very low conception rates.

There are a number of interesting, and, perhaps, unique dietary and nutritional characteristics of the Ngisonyoka Turkana. Low energy intakes seem not to place limits on adult stature nor on women's fertility, which is high, nor on infant mortality, which is moderate (see below). Season variations in body weight and composition reflect frequent hungry periods where both body fat and muscle losses occur. High protein intakes along with physical activity may contribute to recovery of lost muscle mass. Some of these dietary and health issues are discussed below.

### Reproduction

Reproductive biology and fertility are fundamental parameters in understanding the maintenance of population numbers and human labor within the family pastoral unit, and, in fact, in understanding the fundamental social structure of the Turkana. For example, the Turkana are a polygynous society, where men and their families pay brideprice in the form of livestock to acquire wives. Livestock, then, as a form of wealth, contributes not only food for sustenance, but for males, the means to reproduce, as well. This links a herding family's ability to manage and increase its

162    M.A. Little

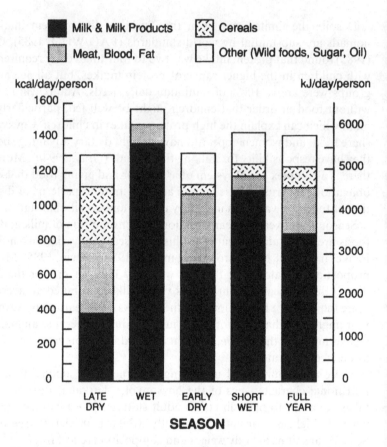

Figure 7.7. Seasonal variations in dietary intake (based on energy intake) from studies by Galvin (1985) of several nomadic Ngisonyoka Turkana families. Data were gathered during a full year (1981–82). When milk is abundant during wet seasons, it is the primary food. During limited milk production by livestock, animals are bled, small stock are slaughtered and there is greater dependence on other foods such as maizemeal and gathered foods.

livestock holdings as an economic endeavor directly to its ability to pay brideprice (livestock) to other families in order to enhance its reproductive capacity. At the same time that a family acquires women through payment of brideprice, it will acquire livestock through payments received for its daughters. These exchanges are negotiated carefully, contribute to the redistribution of wealth in the population, and are essential in the production of children who will serve as the labor needed to herd the animals. A further association with human reproduction is that of livestock reproduction (Leslie and Dyson-Hudson, 1999). When imbalances occur within family herding units between human labor for herding and the numbers of animals and their growth in numbers, then there are institutionalized mechanisms for redistributing animals and human labor through loaning/ borrowing or asking/giving exchanges. Wealth tends to be an ephemeral concept since large herds (wealth in livestock) must be balanced with large numbers of dependants (labor, i.e., wives and children of the herd owner) who draw on the resources of the wealth (milk, blood, meat).

In a polygynous society, such as the Ngisonyoka Turkana, herd owners acquire wives sequentially as they accumulate the livestock needed for brideprice. Hence, the first wife is really a spouse in a monogamous union, and as later wives are acquired, the family unit moves from one to two or more wives. As the family herding unit matures and sons reach marriageable ages (late 20s to early 30s), there may be competition between fathers (herd owners) and sons for the livestock needed to pay brideprice. This probably serves to limit the number of wives that all but the most wealthy herd owners can acquire. Within this extremely complex social, economic, and reproductive system that has been outlined in its barest essentials, Brainard (1991) and, later, Leslie (Leslie *et al.*, 1999*a*) have accumulated substantial information on the reproduction and fertility of the nomadic Turkana. Ngisonyoka women, for example, have a relatively high lifetime fertility of between 6.6 and 7.1 live births (Brainard, 1991; Leslie *et al.*, 1999*a*). By age 65 years, herd owners will have, on average, 2.5 wives and about 10 offspring (Leslie *et al.*, 1999*b*). The fertility rate for women shows considerable fluctuation by season and by year. For example, Leslie and Fry (1989) found a marked seasonality in births that was correlated with rainfall and food availability, and Gray (1992) and Leslie *et al.* (1999*a*) found a doubling of birth rates during a series of years with good rainfall when compared with several years of very poor rainfall. Short-term and long-term environmental fluctuations, combined with the economic risks of keeping livestock, contribute to patterns of reproduction that show considerable variation through time.

### Growth of infants, children, and adolescents

The growth patterns of young people are largely a reflection of their health, well-being, nutritional status, and other circumstances of their lives. Anthropometric surveys to assess growth of nomadic children and adults in body size and composition began in 1981 and were continued sporadically through 1994. Most measurements were reported from cross-sectional studies, but some data were longitudinal (Little *et al.*, 1983, 1993; Little

164　　M.A. Little

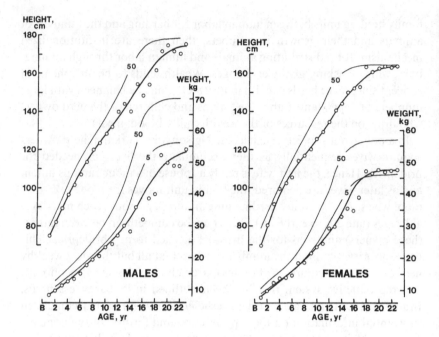

Figure 7.8. Height and weight of Ngisonyoka Turkana youths from infancy to adulthood. Data are from Little et al. (1983); US National Center for Health Statistics (NCHS) reference data are from Hamill *et al.*, (1979).

and Johnson, 1987). Several surveys of settled Turkana schoolchildren were conducted at the irrigation settlements of Nakwamoru and Katilu on the Turkwel River and Morulem on the Kerio River (Little and Gray, 1990; Little *et al.*, 1993).

Height and weight of nomads from birth to adulthood are given in Figure 7.8. Nomadic infants and children have a slow growth pattern and are shorter than United States children (based on National Center for Health Statistics (NCHS) reference values) until adulthood, when Turkana catch up and the two populations become equal in height. The equivalent height is achieved by Turkana continuing to grow into the late teens for women and early 20s for men. Weight is very low by US standards, falling consistently close to or below the US 5th percentile. At all ages from late infancy and beyond, Turkana children and young adults are quite linear or thin in physique, and by world standards Turkana men and women are very tall. Recumbent length of both nomadic and settled infants is close to US norms up to 2 years of age, after which linear growth slows down slightly (see Figure 7.9a). In contrast, infant growth in weight

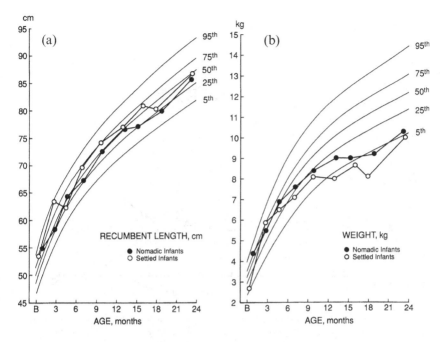

Figure 7.9. Growth of infants during the first two years: (a) recumbent length,
(b) weight. After Little *et al.*, 1993; NCHS data are from Hamill *et al.* (1979).

begins to falter by 6 months of age compared with US reference values,
and drops to about the 5th percentile by 24 months of age (see Figure
7.9b). Other measures of faltering that begin at about 6 months of age are:
head circumference (25th percentile at 24 months of age) and both upper
arm and mid-calf circumferences (10th percentiles at 24 months of age;
Little *et al.*, 1993). Gray (1998) provided evidence that the ability of very
young infants to maintain body weight at around the 50th percentile of
NCHS reference values is a function of Turkana mothers feeding their
infants butterfat during the first 6 months of infants' lives.

As noted, child growth is slow but prolonged and the rapid growth
associated with adolescence is less marked in Turkana boys and girls than
in US adolescents. The peak velocity during adolescence is of low ampli-
tude and it occurs about 2 years later than in US adolescents (see Figure
7.10). Menarche tends to be late by US standards, at about 16.5 years of
age (Gray, 1994*b*).

Comparisons between nomadic and settled children prior to adolescence
suggest that nomadic children are larger than the settled up to about 5

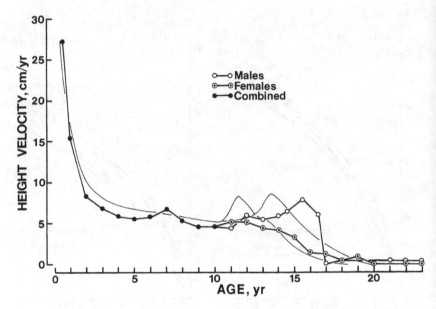

Figure 7.10. Growth velocity curves for Turkana boys and girls (combined up to age 10 years) after Little and Johnson (1987) compared with US children and adolescents from the Denver study (Hansman, 1970).

years of age, after which the settled are larger. This cross-over may result from dietary supplementation of settled children in mission schools, who are provided breakfast and lunch on a daily basis.

Figure 7.11 is a simple model of some of the influences on growth processes from the prenatal period to adulthood. Environmental conditions (rainfall, seasonality, year-to-year variation, disease patterns) influence the livestock productivity, nutritional status and health of the mother, and ultimately prenatal and infant growth of her offspring. The economic status of the herd owner has a direct impact on the mother's health status, and a more direct influence on the health status of the child and adolescent as the child grows and becomes more independent of the mother. The associations are strong between the mother's health status during pregnancy and prenatal growth, and between the mother's health status during the time of breastfeeding and infant growth. These are important relationships partly because of the limited food resources and the fragility of health and nutritional status in women during reproduction and the later period of child rearing. Some of the costs of reproduction in women are described below.

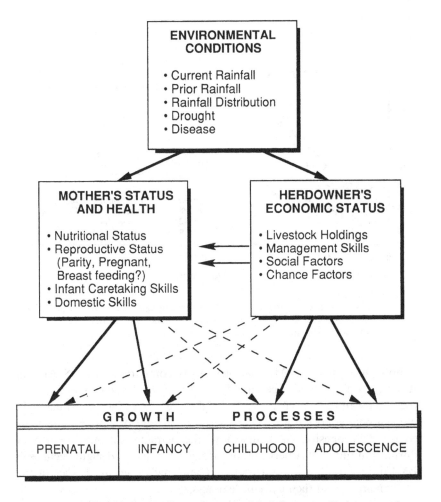

Figure 7.11. A diagram illustrating some of the influences on Turkana growth processes from prenatal growth to adolescence. The solid lines indicate stronger relationships than the interrupted lines. After Little (1995).

### Body composition

Although Turkana are tall as adults, they are very lean with limited fat stores and small muscle masses, although women's arm muscle diameters are nearly as great as men's, presumably because of frequent lifting and carrying heavy objects such as bundles of firewood and containers of water (Little *et al.*, 1983; Little and Johnson, 1986).

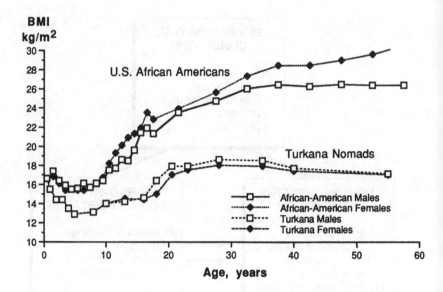

Figure 7.12. Comparison of body mass index (weight/height², kg/m²) of Turkana males and females and US African American males and females. Data on African Americans are from Frisancho (1990).

Body mass index (BMI) of Turkana is compared with US African Americans at different ages in Figure 7.12. The BMI is a measure of robusticity and it is given as weight/height² (kg/m²). In the US, values of 25 units or greater are considered as obese, and 30 units or more are considered as serious obesity. The figure shows the characteristic increase of BMI during adulthood of Americans, particularly women, and slight declines in BMI during adulthood in Turkana. These low values at all ages of the Turkana represent their extreme leanness.

Age changes in body composition are most marked in fat or adipose tissue composition as measured by skinfolds (see Figure 7.13). Fat composition in females shows a steep rise throughout adolescence and then a more gradual decline into adulthood until about age 50 years. Boys show a slight rise into the 20s and then a more gradual decline than women to age 50 years. The decline in female fat composition almost certainly results from the additional energy costs of reproduction to women in light of the limited food energy availability to the Turkana. For example, an average Turkana women has a completed fertility of seven live births (5.25 years of cumulative pregnancy) (Leslie et al., 1988) and 18 months of breastfeeding for each of the seven infants (10.5 years of cumulative lactation; Gray, 1994b). This amounts to her food energy needs being substantially elevated

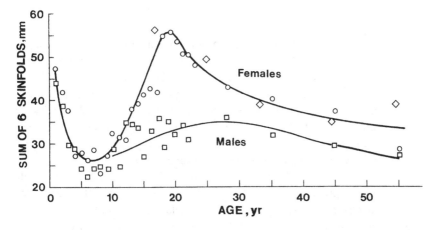

Figure 7.13. Variations by age and sex in the sum of six skinfolds for nomadic Turkana.

for about 15 years, or half of her reproductive life. Tests were made to determine if the elevated energy requirements were associated with reproduction (pregnancy and lactation) or age changes (Little *et al.*, 1992). Relationships between skinfolds and parity were stronger than between skinfolds and age for both nomadic and settled Turkana women, although the relationships were strongest in women who had had less than five or six live births. Also, there is some evidence now (Pike, 1996) that birth weights of late parity nomadic infants are low, suggesting a causal relationship between maternal body composition and reproductive health. That is, late parity mothers with depleted fat reserves have lower birth weight infants who are at greater risk of mortality.

Another issue in body composition, seasonal change, is superimposed on the age changes linked to reproduction. With the seasonal rainfall pulse in this semi-arid environment, several intersecting cycles appear: (1) rainfall, with the several-month *long rains* and late-year *short rains*; (2) the vegetation flush, which shows a slight lag behind the rains; (3) the livestock recovery of "condition" or body weight which follows the vegetation flush; and (4) the recovery of human body weight that follows the increased milk production of the livestock. There is considerable evidence for seasonal variation in body composition and weight for Turkana children at all ages and adults (Galvin, 1985; Galvin and Little, 1987; Little *et al.*, 1993; Shell-Duncan, 1995). If body composition reflects health and nutritional status, then season of the year is an important variable affecting health, reproduction, and other aspects of human well-being.

### Work and activity

Human physical activity is what drives the Turkana population. Livestock provide food (milk, meat, blood) and other products (hides, traded food, other items), but animals must be cared for and managed, and labor is required to harvest the products of the animals for human consumption (milking, slaughtering, bleeding, hide preparation). Physical activity or work influences dietary requirements, such that high levels of physical activity require high intakes of food calories (energy). Labor requirements also are linked to reproduction and child growth. Low fertility of a herd owner's wives, or high infant mortality, or prolonged illness during childhood, or poor growth progress will limit the development of the labor pool to be drawn on for herd management and place additional work requirements on a smaller labor pool to manage the livestock. Strategies of reproduction are closely tied to labor needs and to complex aspects of the social structure of Turkana society. Other factors that influence work are the traditional divisions of labor by age and by sex, and the seasonal changes in work requirements associated with watering and seeking appropriate forage for different species of livestock. Disease also limits the ability to perform work, and individuals will respond to inadequate food energy intake by reducing physical activity to become more sedentary.

A diagram illustrating relationships among physical activity variables of the Ngisonyoka Turkana labor force is shown in Figure 7.14. The variable environment and social processes structure health, developmental, and other influences on the immediate factors of health, body composition, and physical (physiological) fitness. The physical (work) activities of the labor force (men, women, children) are structured by age and sex, and include both sedentary and more vigorous activities. Some examples can be given of results of studies of physical activity.

Women's activities are highly varied and structured according to their age, position in the household (e.g., senior or junior wife), responsibilities for infants or children, and whether the labor force is adequate for herding the household's livestock. For example, senior wives are less active because childbearing may be completed, and younger wives of the herd owner assume responsibilities for many household tasks. An observation that is an indicator of the extent of physical work conducted by women is their upper body muscle mass and strength (Little et al., 1983; Little and Johnson, 1986). Studies of upper arm and forearm cross-sectional muscle and bone areas indicated that Turkana women have lean tissue areas that are equivalent to Turkana men, who are 7% heavier and 4% taller than the women. In addition, although Turkana women are about 20% lighter in

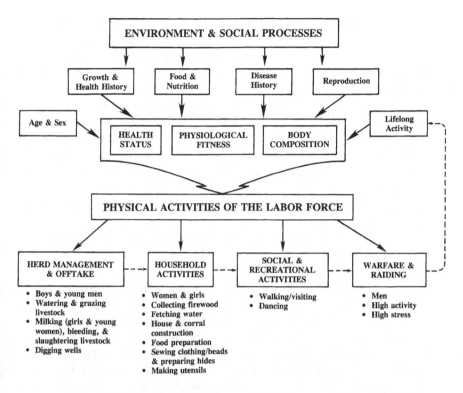

Figure 7.14. A diagram illustrating some of the influences on Turkana physical activity processes. The overarching influences arise from the environmental variation of the dry savanna and the social and behavioral processes of Turkana culture.

body weight than their US age mates, upper arm and forearm lean tissue areas are slightly greater in Turkana than US young women. Moreover, Turkana women show very little losses with age in either arm strength or arm muscle. These measures of muscle size and strength in Turkana women are almost certainly a function of socially prescribed women's activities (see Figure 7.14). Lifting heavy loads and arm exercise character-ize much of the work done by women. Milking large numbers of animals will develop the grip flexor muscles; carrying water (up to 20 liters) and firewood on the head will develop arm muscles; house and corral construc-tion require lifting and carrying; and young women must assist in raising water from wells in basins to provide substantial amounts of water for livestock during the dry season. This is a labor intensive activity in which both young women and young men will participate. Herd owners (gen-erally age > 30 years) are more sedentary than young men or women of all ages, but they do walk considerable distances. Young men (generally age < 30 years) and boys, on the other hand, have immediate responsibility for animal grazing and watering. Distances traveled by herders during the dry season may be up to 20 km/day and require considerable work to manage the livestock particularly during drought (Coppock *et al.*, 1986*b*; McCabe, 1994). Total distances traveled by herd boys or young men in a given year or season may be as much as 800 km (McCabe, 1990). Curran-Everett (1994) found that exercise capacity ($VO_2$ max) in men was associated with work history (active herders had higher fitness than former herders or non-herders). In addition, declines in exercise capacity that are characteristic of Western men do not show up in the Turkana nomads. In other words, levels of physical fitness in Turkana men do not decline with age, at least up to age 55 years.

Nomadic Turkana children are quite active, physically. Boys will begin herding small flocks of young goats or sheep near the home settlement by age 5 years, and young girls will assist with household tasks by about the same age. In children, the principal demands on dietary energy are basal metabolism, physical activity, and growth. The greatest differences among Turkana children in diets, activity, and growth and body composition are between the Ngisonyoka nomads and the settled Turkana who can be found at the towns of Katilu, Morulem, and Nakwamoru. Table 7.1 lists some of the differences between the two groups of Turkana children.

The inconsistent results of these comparisons are the higher caloric intake of the nomads than the settled children and the larger sizes at all ages (4–9 years) of the settled than the nomadic children. A resolution of these inconsistencies probably rests on: (1) the reduced physical activity of the settled children because of their long hours in school each day and the corresponding reduced energy requirements (from lower energy expenditure) for the schoolchildren, and (2) the unknown additional food taken in by the schoolchildren in the home or elsewhere. When the settled schoolchildren were questioned about their diets in addition to the school food, they uniformly stated that no other food was eaten. However, it is almost certain that the school food (on which caloric intakes were calculated) was only part of the total daily intake. Since, settled children aged 1 to 4 years (prior to attending school) were generally *smaller* than nomadic children of the same ages (Brainard, 1990; McCarthy, 1980), then it is likely that the school food supplementation and limited physical activity of schoolchildren have given the older settled children (aged 4 to 9 years) an advantage in growth processes.

Table 7.1. *Comparisons of settled and nomadic children aged 4 to 9 years according to diet, activity, growth and body composition*

|  | Settled Turkana children | Nomadic Turkana children |
| --- | --- | --- |
| Diet and nutrition | Breakfast and lunch (maize porridge with milk) provided by schools. Not known what food is available at home. Based on school foods, average caloric intake was 1000 kcal/day and protein intake was 35 g/day | High-protein food (milk) when available. Adults will undergo deprivation to reserve milk for children (children have high priority for food). Based on Galvin's (1985) survey calories were 1350 kcal/day and protein intake was 73 g/day |
| Physical acitivity | Children spent about 5 hours per day in the classroom in sedentary activity | Children were physically very active: playing, herding, or carrying out household tasks |
| Growth patterns | Between ages 4 and 9 years, settled children were taller and weighed more than nomads | Arm circumferences were equal in settled and nomadic children; leg circumferences were smaller in the nomads than the settled |
| Body composition | Settled children were more robust than nomads. Summed skinfolds were greater than in nomads, but differences were minimal for upper arm. Girls' skinfolds were greater than boys' | Nomadic girls had larger fat deposits (skinfolds) than boys, but skinfolds in girls were smaller than settled boys |

From: Brainard (1991), Little and Gray (1990), Galvin (1985).

### Health and morbidity

Characterizing the health of the Ngisonyoka Turkana nomads is a difficult task, because of substantial gaps in our knowledge and some inconsistencies in what we do know. Jean Brainard (1981) gathered data on health of long-settled and recently settled Turkana at the Nakwamoru settlement, and Karen Shelley (1985) conducted the earliest survey of health beliefs, folk medicine, and estimates of disease prevalence for the Ngisonyoka. Nomads have limited access to Western medical assistance, although there are clinics staffed by trained nurses scattered throughout Ngisonyoka territory. The clinics, which are located in towns (Lokichar, Lokori,

Katilu, Nakwamoru, Morulem) and often associated with missions, are seldom attended by members of nomadic families. In the nomadic population, most health care, then, is provided by women in the context of the family. Sickness is perceived by Turkana as the result of "misfortune" and dealt with by "resolving causes and negative effects of misfortune brought about by harmful agents in the environment" (Shelley, 1985, p. 252).

Brainard (1981, 1991) worked with Turkana at the Nakwamoru settlement and irrigation scheme. She gathered data on health and fertility of the Turkana living at the settlement, drawing on interview data as well as records from patients who attended the Nakwamoru clinic. Nakwamoru residents were divided for analysis into those who had spent most of their lives as nomadic pastoralists and were settled recently, and those whose lives had been primarily spent as settled cultivators (Ngikatak Turkana). Based on clinic records, Brainard found that major causes of morbidity at Nakwamoru were malaria, acute respiratory infections (ARI), eye infections/conjunctivitis (primarily *Chlamydia trachomatis* infection), and gastrointestinal (GI)/diarrheal infections. About 50% of all clinic admissions were diagnosed as malaria, ARI, and GI/diarrheal infections. Nakwamoru had a particularly high prevalence of malaria because of its proximity to the Turkwel River forest. The high prevalence of ARI (21%) and GI/diarrheal (8.4%) infections were attributed to poor sanitation and the crowded conditions of the nucleated settlement. Gray's (1998) study of nomadic Ngisonyoka nursling infants also demonstrated the high prevalence of respiratory infections (including whooping cough and bronchitis) and malarial symptoms in infants, but relatively few diarrheal infections.

The central interest of Shelley (1985) was on the nomadic Turkana healthcare system and beliefs as part of a wider cultural system. However, she also did clinical assessment of 111 Ngisonyoka adults and children with the assistance of a clinical intern, and collected stool specimens from 36 children. She reported that malaria was the most common cause of sickness and that spleen and liver enlargement and pain were common complaints. Liver enlargement was most prevalent in children aged 5–14 years. Stool specimens, which were analyzed at the Kenyatta University Hospital Laboratory in Nairobi indicated that only 4 of the 36 children tested had evidence of parasites. Two children had *Giardia lamblia* and two had *Chilomastix mesnili*. Although she took no measurements to assess nutritional status, Shelley (1985) observed that there were no clinical signs of marasmus and kwashiorkor, and few signs of other nutritional deficiencies. Gray's (1992, 1998) studies supported those of Shelley. And it appears that nomads are less at risk of diarrheal disease than are settled infants and children.

Infant mortality (deaths per 1000 births during the first year) and child mortality can be used as measures of health in a population. Brainard (1986) found an infant mortality rate for permanently settled women at Nakwamoru of 133/1000 and a rate for recently settled women 30 years of age or older of 243/1000, nearly double the permanently settled rate. The recently settled women had lived a part or all of their reproductive lives in the nomadic pastoral sector; hence, Brainard concluded that Turkana pastoralists were likely to have higher infant mortality than settled Turkana. Leslie *et al.* (1999*b*) estimated infant mortality in Ngisonyoka pastoralists at between 90 and 140 deaths per 1000. These infant mortality values are lower than Brainard's (1986), where the higher infant mortality of the recently settled Turkana may reflect impoverishment of the women due to failed pastoral activities. Leslie *et al.* (1999*b*, p. 538) note that "despite the considerable challenges posed by their environment, the Ngisonyoka mortality rates are low to moderate compared to those of other African pastoralists for whom data are available." All of these mortality values must be interpreted with some degree of caution since these data are largely recall and drawn from interviews. Moreover, there is likely to be substantial variation in mortality from year to year because of frequent droughts and other disasters that influence health of the nomadic pastoralists.

The most detailed studies of infectious disease in children were conducted by Shell-Duncan (1994, 1995; Shell-Duncan and Wood, 1997) who measured cellular immunity by an antigenic challenge known as delayed-type hypersensitivity (DTH). She tested 57 Ngisonyoka Turkana children during the rainy season and then 5 months later during the dry season, with a skin test of five antigens and a control. The sizes of the indurations were measured after 48 hours. She found that, when compared with US and Peruvian children, the Turkana children showed minimal positive skin induration responses (56.2% were nonreactive), and concluded that the Turkana children were anergic and immunosuppressed, largely as the result of widespread malnutrition (Shell-Duncan, 1994). In a later work (Shell-Duncan, 1995), no seasonal variation in the DTH test was found and it was concluded that the nutritional stress experienced by pastoral nomadic children was of sufficient magnitude to produce immunosuppression throughout the year despite seasonal variation in food availability, particularly when infection rates were higher in the (rainy) season when food was more abundant. Most recently (Shell-Duncan and Wood, 1997), close associations were found between DTH responsiveness and acute respiratory and diarrheal infections.

There are several caveats to be raised about the conclusions drawn from this work. First, nomadic Turkana children can be depicted as lean in physique by Western standards, which is largely the result of limited food energy (Little *et al.*, 1983; Galvin, 1985). On the other hand, Ngisonyoka Turkana children show hypersufficiency for protein intake; hence, limited protein, which may be a cause of immune deficiency in other populations, is certainly not a health threat for the Turkana. I believe it misrepresents the status of Turkana children to characterize them as chronically mal-nourished, particularly when only a minority of children are identified by US anthropometric standards as wasted (26% were low weight-for-age) or stunted (15% were low height-for-age; Shell-Duncan, 1995). Second, as members of a population that is relatively isolated from settled groups and does not live under congested, nucleated conditions, Turkana children may not respond to the DTH challenge because they have not been exposed to the microorganisms that were applied. In the case of the tuberculin antigen, 32% responded with positive induration, indicating immunocompetence for the sole antigen where the disease (TB) is known to be prevalent in the Turkana population from clinical records. Third, although a quarter of the children were assessed as chronically mal-nourished at mild to moderate levels, more than half of the children were identified as immunosuppressed (Shell-Duncan, 1995). Finally, about 47% of the children who were positive for the DTH test in the wet season, then tested negative during the dry season. Hence, children were being moved between categories of immunoincompetent and immunocompetent over a period of about 5 of 6 months.

One of my concerns about the characterization of Turkana children as *chronically malnourished*, and hence, *immunosuppressed*, is that there is considerable variation in the health status of the nomadic children. The majority of children are remarkably healthy despite the absence of West-ern health intervention, and considering the conditions of food availability and hygiene in the children's lives. Infant and child mortality rates are high compared with the well-off residents of Western nations, but not as high as mortality in other pastoral populations (Leslie *et al.*, 1999*b*), and probably not as high as in settled Turkana populations. The settled Turkana carry heavy malaria burdens and are unable to maintain good nutritional health in children, so must depend on school food supplementation for the children. Another concern, bears on the disruptive effects of outside assist-ance to the Turkana. If they are characterized as chronically malnourished and immunoincompetent and subject to high infection rates, they will be identified as impoverished and in need of outside intervention. Such inter-vention in the past has contributed to the establishment of famine-relief camps that have moved Turkana from the pastoral sector to a state of dependency lasting for a generation or more. One would hope to avoid this in the future.

S.J. Gray (personal communication) has hypothesized that repeated efforts to settle the Karimojong of Uganda (a population closely related to the Turkana) have had a negative impact on human health and adaptability in that population. Preliminary results from recent fieldwork suggest much higher levels of mild to moderate, as well as, acute, malnutrition in Karimojong children than among nomadic Turkana. Malnutrition in Karimojong children appears to have less to do with energy and micronutrient deficiencies than with chronic diarrheal, acute respiratory, and malarial infections in the under–5 years population. Preliminary results also indicate that infant and child mortality rates are markedly higher in the Karimojong than in the nomadic Turkana. All of this suggests the operation of complex interactions between nutritional status and acute infections in infants and children (Martorell, 1980), particularly in populations such as the Karimojong that have undergone decades of social and economic change and health disruption.

### Discussion

Some of the basic attributes of the Ngisonyoka Turkana system that were discussed at the beginning of the chapter are listed below:

- the environment is semi-arid with limited resources;
- there is substantial environmental variability (both temporal and spatial);
- livestock convert vegetation that is unpalatable to people to products (milk, meat, blood) that can be consumed by humans;
- pastoral nomadism, as a primary means of environmental exploitation, is opportunistic;
- opportunistic exploitation is possible through mobility of family herding units and livestock;
- the people have a strong attachment to their animals with a sophisticated knowledge of livestock and livestock management;
- there is a complex social system of exchange and reciprocity that maintains balances between livestock numbers and human labor and facilitates survival during difficult times;
- the people attempt to maximize livestock holdings and human numbers;

178     M.A. Little

- there is a relatively high failure rate and loss of people from the pastoral sector, particularly during prolonged drought;
- the arid savanna ecosystem tolerates wide fluctuations in livestock and human populations;
- the multiple perturbations in this system result from external forces such as rainfall;
- the Turkana pastoralists are highly responsive to externally driven environmental variations in their fertility, mortality, out-migration, growth, activity patterns, subsistence patterns, and general social behavior.

In addition to variation in pattern and intensity of rainfall and drought as external driving forces, there are many social, economic, and larger-scale governmental factors that affect the Turkana population. There are intersections of inter-ethnic raiding and warfare, governmental and non-governmental intrusion, provision of food relief, and periodic appearance of livestock disease. For example, between 1976 and 1995, during the duration of our studies, there were three droughts, many episodes of inter-ethnic raiding (Pokot/Turkana), several livestock epidemics, and two major Government of Kenya military interventions. These events were superimposed on seasonal patterns of subsistence and culture that made prediction difficult and underline the need for long-term comprehensive investigations of pastoral peoples.

Some of these issues are discussed at greater length in the context of the Turkana ecosystem as a nonequilibrial system by Gray *et al.* (2002).

As noted, the Turkana are highly responsive to environmental variation in their human biology, health, and adaptability. Female fertility is high during good years (abundant food resources) and lower during bad years (low rainfall/drought) (Gray, 1992; Leslie *et al.*, 1999a); births tend to be patterned according to seasonality (Leslie and Fry, 1989); and a slow and prolonged growth of children and adolescents allows individuals to achieve a tall stature as adults despite limited food energy intakes (Little *et al.*, 1983; Little and Johnson, 1987). Here, then, are examples of further opportunistic responses of a biobehavioral nature. Behavioral opportunistic or contingent adaptive responses (Dyson-Hudson, 1980) in movement and livestock management can be carried out quickly. Biobehavioral opportunistic or adaptive responses in reproduction, growth, and population adjustment require a longer time frame, and include responses that are often difficult to perceive.

It is only through multidisciplinary, environmentally integrated, bio-behavioral, and long-term investigations that the detailed character of human populations and their adaptive mechanisms can be uncovered. The Turkana research reflects some degree of progress in these directions.

## References

Brainard, J.M. (1981). Herders to farmers: the effects of settlement on the demography of the Turkana population of Kenya. PhD dissertation in anthropology, State University of New York, Binghamton.

Brainard, J.M. (1986). Differential mortality in Turkana agriculturalists and pastoralists. *American Journal of Physical Anthropology* **70**, 525–536.

Brainard, J.M. (1990). Nutritional status and morbidity on an irrigation project in Turkana District, Kenya. *American Journal of Human Biology* **2**, 153–163.

Brainard, J.M. (1991). *Health and Development in a Rural Kenyan Community.* New York: Peter Lang Publishing.

Coppock, D.L., Ellis, J.E. and Swift, D.M. (1986*a*). Livestock feeding ecology and resource utilization in a nomadic pastoral ecosystem. *Journal of Applied Ecology* **23**, 573–583.

Coppock, D.L., Swift, D.M., Ellis, J.E. and Galvin, K. (1986*b*). Seasonal patterns of energy allocation to basal metabolism, activity and production for livestock in a nomadic pastoral ecosystem. *Journal of Agricultural Science* **107**, 357–365.

Coughenour, M.B., Ellis, J.E., Swift, D.M., Coppock, D.L. Galvin, K., McCabe, J.T. and Hart, T.C. (1985). Energy extraction and use in a nomadic pastoral ecosystem. *Science* **230**, 619–625.

Coughenour, M.B., Coppock, D.L., Rowland, M. and Ellis, J.E. (1990*a*). Dwarf shrub ecology in Kenya's arid zone: *Indigofera spinosa* as a key forage resource. *Journal of Arid Environments* **18**, 301–312.

Coughenour, M.B, Coppock, D.L. and Ellis, J.E. (1990*b*). Herbaceous forage variability in an arid pastoral region of Kenya: importance of topographic and rainfall gradients. *Journal of Arid Environments* **19**, 147–159.

Curran-Everett, L.S. (1990). Age, sex, and seasonal differences in the work capacity of nomadic Ngisonyoka Turkana pastoralists. PhD dissertation in Anthropology, State University of New York, Binghamton.

Curran-Everett, L.S. (1994). Accordance between $VO_2$ max and behavior in Ngisonyoka Turkana. *American Journal of Human Biology* **6**, 761–771.

Dyson-Hudson, N. (1973). Turkana. In *Primitive Worlds*, pp. 84–111. Washington, DC: National Geographic Society.

Dyson-Hudson, N. (1980). Strategies of resource exploitation among East African savanna pastoralists. In *Human Ecology in Savanna Environments*, ed. D.R. Harris, pp.171–184. London: Academic Press.

Dyson-Hudson, R. (1989). Ecological influences on systems of food production and social organization of South Turkana pastoralists. In *Comparative Socioecology: The Behavioral Ecology of Humans and Other Mammals*, ed. V. Standen and R.A. Foley, pp. 165–193. Oxford: Blackwell.

Dyson-Hudson, R. (1999). Turkana in time perspective. In *Turkana Herders of the Dry Savanna: Ecology and Biobehavioral Response of Nomads to an Uncertain Environment*, ed. M.A. Little and P.W. Leslie, pp. 24–40. Oxford: Oxford University Press.

Dyson-Hudson, R. and Dyson-Hudson, N. (1980). Nomadic pastoralism. *Annual Review of Anthropology* 9, 15–61.

FAO. (1967). *FAO Production Yearbook*. Rome: Food and Agricultural Organization.

FAO/WHO. (1985). *Energy and Protein Requirements*. WHO technical Reports Series No. 724. Geneva: Food and Agricultural Organization/World Health Organization.

Frisancho, A.R. (1990). *Anthropometric Standards for the Assessment of Growth and Nutritional Status*. Ann Arbor: University of Michigan Press.

Galvin, K. (1985). Food procurement, diet, activities and nutrition of Ngisonyoka, Turkana pastoralists in an ecological and social context. PhD dissertation in anthropology, State University of New York, Binghamton.

Galvin, K.A. (1992). Nutritional ecology of pastoralists in dry tropical Africa. *American Journal of Human Biology* 4, 209–221.

Galvin, K. and Little, M.A. (1987). Seasonal patterns of body size and composition among pastoral nomads from northwest Kenya (abstract). *American Journal of Physical Anthropology* 72, 200.

Galvin, K.A., Coppock, D.L. and Leslie, P.W. (1994). Diet, nutrition, and the pastoral strategy. In *African Pastoralist Systems: An Integrated Approach*, ed. E. Fratkin, K.A. Galvin and E.A. Roth, pp. 113–132. Boulder, CO: Lynne Rienner.

Gray, S., Leslie, P. and Akol, H.A. (2002). Uncertain disaster: Environmental instability, colonial policy, and resilience of East African pastoralist systems. In *Human Biology of Pastoral Populations*, ed. W.R. Leonard and M.H. Crawford, pp. 99–130. Cambridge: Cambridge University Press.

Gray, S.J. (1992). Infant care and feeding among nomadic Turkana pastoralists: implications for child survival and fertility. PhD dissertation in anthropology, State University of New York, Binghamton.

Gray, S.J. (1994a). Correlates of dietary intake of lactating women in South Turkana. *American Journal of Human Biology* 6, 369–383.

Gray, S.J. (1994b). Comparison of effects of breast-feeding practices on birth-spacing in three societies: nomadic Turkana, Gainj, and Quechua. *Journal of Biosocial Science* 26, 69–90.

Gray, S.J. (1998). Butterfat feeding in early infancy in African populations: new hypotheses. *American Journal of Human Biology* 10, 163–178.

Gulliver, P.H. (1955). *The Family Herds: A Study of Two Pastoral Tribes in East Africa, the Jie and the Turkana*. London: Routledge and Kegan Paul.

Hamill, P.V.V., Drizd, T.A., Johnson, C.L., Reed, R.B., Roche, A.F. and Moore, W.M. (1979). Physical growth: National Center for Health Statistics percentiles. *American Journal of Clinical Nutrition* 32, 607–629.

Hansman, C. (1970). Anthropometry and related data: anthropometry, skinfold measurements. In *Human Growth and Development*, ed. R.W. McCammon, pp. 101–154. Springfield, IL: C.C. Thomas.

Lamphear, J. (1988). The people of the grey bull: the origin and expansion of the Turkana. *Journal of African History* **29**, 27–39.

Leslie, P.W. and Dyson-Hudson, R. (1999). People and herds. In *Turkana Herders of the Dry Savanna: Ecology and Biobehavioral Response of Nomads to an Uncertain Environment*. ed. M.A. Little and P.W. Leslie, pp. 232–247. Oxford: Oxford University Press.

Leslie, P.W. and Fry, P.H. (1989). Extreme seasonality of births among nomadic Turkana pastoralists. *American Journal of Physical Anthropology* **79**, 103–115.

Leslie, P.W, Bindon, J.R. and Baker, P.T. (1984). Caloric requirements of human populations: a model. *Human Ecology* **12**, 137–162.

Leslie, P.W., Fry, P.H., Galvin, K. and McCabe, J.T. (1988). Biological, behavioral and ecological influences on fertility in Turkana. In *Arid Lands Today and Tomorrow: Proceedings of An International Research and Development Conference*. ed. E.E. Whitehead, C.F. Hutchinson, B.N. Timmermann and R.C. Varady, pp. 705–712. Boulder, CO: Westview Press.

Leslie, P.W., Campbell, K.L., Campbell, B.C., Kigondu, C.S. and Kirumbi, L.W. (1999*a*). Fecundity and fertility. In *Turkana Herders of the Dry Savanna: Ecology and Biobehavioral Response of Nomads to an Uncertain Environment*, ed. M.A. Little and P.W. Leslie, pp. 248–278. Oxford: Oxford University Press.

Leslie, P.W., Dyson-Hudson, R. and Fry P.H. (1999*b*). Population replacement and persistence. In *Turkana Herders of the Dry Savanna: Ecology and Biobehavioral Response of Nomads to an Uncertain Environment*, ed. M.A. Little and P.W. Leslie, pp. 280–301. Oxford: Oxford University Press.

Little, M.A. (1995). Growth and development of Turkana pastoralists. In *Research Frontiers in Anthropology: Advances in Archaeology and Physical Anthropology*, ed. P.N. Peregrine, C.R. Ember and M. Ember. Englewood Cliffs, NJ: Prentice-Hall.

Little, M.A. and Gray, S.J. (1990). Growth of young nomadic and settled Turkana children. *Medical Anthropology Quarterly* **4**, 296–314.

Little, M.A. and Johnson, B.R., Jr (1986). Grip strength, muscle fatigue and body composition in Nomadic Turkana pastoralists. *American Journal of Physical Anthropology* **69**, 335–344.

Little, M.A. and Johnson, B.R., Jr (1987). Mixed longitudinal growth of Turkana pastoralists. *Human Biology* **59**, 695–707.

Little, M.A. and Leslie, P.W., ed. (1990). *The South Turkana Ecosystem Project*. Binghamton, NY and Fort Collins, CO: Report to the Government of Kenya, Office of the President.

Little, M.A. and Leslie, P.W., ed. (1999). *Turkana Herders of the Dry Savanna: Ecology and Biobehavioral Response of Nomads to an Uncertain Environment*. Oxford: Oxford University Press.

Little, M.A., Galvin, K. and Mugambi, M. (1983). Cross-sectional growth of nomadic Turkana pastoralists. *Human Biology* **55**, 811–883.

Little M.A., Galvin, K. and Leslie, P.W. (1988). Health and energy requirements of nomadic Turkana pastoralists. In *Coping with Uncertainty in Food Supply*, ed.

I. de Garine and G.A. Harrison, pp. 288–315. Oxford: Oxford University Press.

Little M.A., Leslie, P.W. and Campbell, K.L. (1992). Energy reserves and parity of nomadic and settled Turkana women. *American Journal of Human Biology* **4**, 729–738.

Little, M.A., Gray, S.J. and Leslie, P.W. (1993). Growth of nomadic and settled Turkana infants of northwest Kenya. *American Journal of Physical Anthropology* **92**, 273–289.

Martorell, R. (1980). Interrelationships between diet, infectious disease, and nutritional status. In *Social and Biological Predictors of Nutritional Status, Physical Growth, and Neurological Development*, ed. L.S. Greene and F.E. Johnston, pp. 81–106. New York: Academic Press.

McCabe, J.T. (1990). Turkana pastoralism: a case against the Tragedy of the Commons. *Human Ecology* **18**, 81–103.

McCabe, J.T. (1994). Mobility and land use among African pastoralists: old conceptual problems and new interpretations. In *African Pastoralist Systems: An Integrated Approach*. ed. E. Fratkin, K.A. Galvin and E. A. Roth, pp. 69–89. Boulder CO: Lynne Rienner.

McCarthy, F.D., ed. (1980). The Katilu Irrigation Scheme in Turkana District, Kenya: a socioeconomic–nutrition analysis report of a survey, March 1979. Food and Agricultural Organization (FAO) and Ministry of Economic Planning and Community Affairs. Nairobi: FAO.

Pike, I.L. (1996). Determinants of pregnancy outcome for nomadic Turkana women of Kenya. PhD dissertation in anthropology, State University of New York, Binghamton.

Shell-Duncan, B. (1994). Determinants of infant and child morbidity among nomadic Turkana pastoralists of northwest Kenya. PhD dissertation in anthropology, Pennsylvania State University, University Park.

Shell-Duncan, B. (1995). Impact of seasonal variation in food availability and disease stress on the health status of nomadic Turkana children: a longitudinal analysis of morbidity, immunity, and nutritional status. *American Journal of Human Biology* **7**, 339–355.

Shell-Duncan, B. and Wood, J.W. (1997). The evaluation of delayed-type hypersensitivity responsiveness and nutritional status as predictors of gastrointestinal and acute respiratory infection: a prospective field study among traditional nomadic Kenyan children. *Journal of Tropical Pediatrics* **43**, 25–32.

Shelley, K. (1985). Medicines for misfortune: diagnosis and health care among southern Turkana pastoralists of Kenya. PhD dissertation in anthropology, University of North Carolina, Chapel Hill.

# 8 Economic stratification and health among the Herero of Botswana

RENEE L. PENNINGTON

## Introduction

Differences in child health and survival have been identified in a number of peoples but the causes are not well understood. Comparisons of child survival rates between ethnic groups living side by side suggest that cultural differences have measurable affects on the health of children (Hill *et al.*, 1983; Pennnington and Harpending, 1993). Within these communities large differences in child survival may also be observed, suggesting that differences in parenting or the inter-community distribution of resources influence the health of children. Identifying the causes of differences has clinical and theoretical applications. Accurate identification of children at elevated risk of morbidity may improve the design of intervention strategies and the allocation of limited resources. Child mortality directly affects the reproductive success of parents, and, therefore, variables correlated with child health may be associated with different reproductive strategies of parents.

This chapter examines the relationship between economic differences among Herero and Mbanderu (henceforth collectively referred to as Herero), two groups of closely related rural African cattle people, to growth of children. There are substantial differences among Herero families in the survival of children and an apparent preference for female children that are not observed in neighboring !Kung Bushmen (Pennington and Harpending, 1993). Although both groups live in the northern Kalahari Desert of Botswana and sometimes intermarry, they differ genetically, historically, and culturally. Most conspicuously, as a group, the Herero are prosperous while most !Kung are indigent, perhaps accounting for the difference in child survival patterns (Pennington and Harpending, 1993).

Within the Herero we have observed differences in wealth between households, and in this chapter I discuss the link between livestock holdings (their own measure of wealth) on the health of Herero children as the

183

likely cause of previously observed heterogeneity in child mortality. I also examine the anthropometric status of Herero children in a regional context. This study adds to the sparse literature on the growth and health of African pastoralists.

## Purpose

The survival rate of children in a society may be used as a summary statistic of that society's standard of living. Although a newborn's chances of dying decline until about puberty across populations, the rate of decline varies considerably. Mortality in the period of highest risk, from birth through weaning, is especially sensitive to care-dependent and care-independent factors (Harpending et al., 1990). That is, child care behaviors appear to contribute in important ways to a child's chances of survival as do factors such as local disease ecologies beyond the control of parents. Perhaps the most important correlate of risk is diet, both quality and quantity.

Where death rates of children are high, quality resources are scarce or poorly distributed. Diets deficient in quality or quantity produce individuals with poor growth, inadequate energy reserves and, probably, flawed immune functioning. Consequently, indicators of nutritional status such as anthropometric measures of growth provide an objective means of evaluating the health of individuals and populations, and these must be directly rated to the demography of a region.

### Previous findings

We test here the association between resource holdings and anthropometric indicators of health of two groups of Herero pastoralists. We have documented differences in child mortality among families and an apparent preference for daughters (Pennington and Harpending, 1993) and have observed economic differences. Mortality during infancy (the proportion of newborns who fail to survive from birth to age 1 year) ranged from 0.16 to 0.03. Mortality during early childhood (the proportion of 1-year-olds who fail to survive to age 5 years) ranged from 0.10 to 0.02. The rates are low among Africans but within the ranges reported in Botswana. Mortality among boys has been two to three times that of girls in the past several decades. The excessive mortality rate of boys, which far exceeds expected differences in mortality between the sexes, is indicative of a preference for female children. There are also apparent differences among Herero mothers in that some mothers have much higher death rates among their children than others.

We have been unable to fully explain the apparent sex preference of children and heterogeneity in mortality among mothers in the survival of progeny. They are absent among neighboring !Kung Bushmen (Pennington and Harpending, 1988, 1993). These former hunter-gatherers began settling among Herero and other Bantu cattle ranchers at increasing rates from the 1950s. Dramatic decreases in child mortality, to levels similar to Herero, were concomitant with the switch to more sedentary living. We have attributed the decline to improved weaning diets.

The absence of heterogeneity or apparent sex preference in !Kung but present in Herero suggests that important cultural differences contribute to mortality risks among children. The salient characteristic distinguishing the two groups is the relative prosperity of Herero. Most !Kung remain reluctant participants in the Botswanan economy while nearly all adult Herero own and market cattle. The size of cattle holdings is correlated with all measures of economic well-being in the agricultural based economy of Botswana (Litschauer and Kelly, 1981).

The goal of this study was to assess the contribution of heterogeneity among Herero families in cattle holdings as a cause of heterogeneity in child mortality. We examined health of children using standard anthropometric measures and their relationship to household wealth. We expected to find superior economic success associated with superior indicators of child health. We test this hypothesis below and discuss the potential contribution of heterogeneity in wealth to heterogeneity in child mortality.


### Methods

This study is based on cross-sectional analysis of anthropometric data and censuses of homesteads and livestock herds of Herero living in remote regions of the Ngamiland District of northwestern Botswana. Henry Harpending and I collected these data in May–August 1992. We collected weights, heights, triceps skinfolds, and upper arm circumferences on 828 Herero and recorded reproductive histories, residential information and cattle and small stock holdings from members of 28 homesteads living in three distinct regions of Ngamiland. This paper reports on correlations between socioeconomic differences, as measured in livestock, and anthropometric indicators of health in children living in these communities. To preserve the economic privacy of Herero households, I do not give details about sizes of individual herds or specifically identify the families and regions we visited.

### Ethnographic background

The Herero in this study are Bantu-speaking pastoralists who occupy scattered homesteads in the northern Kalahari Desert. I use the term Herero to refer to the two groups of Herero-speaking peoples in Botswana, the Herero proper and the closely related Mbanderu. Distinctions between the two groups are largely political and of historical origin. They share customs, intermarry, and stand out among other groups in Botswana due to the long, gaudy dresses and head pieces, copied after turn-of-the-century missionaries in Southwest Africa (now Namibia), worn by women.

The number of Herero living in Botswana is unknown, though there were probably fewer than 20 000 of them, mostly living in the Ngamiland region, when this study took place in 1992. Many more Herero live in Namibia, and most of the Herero in Botswana today are descended from refugees who fled that region during the Herero-German War of 1904. Many Herero proper consider Namibia their homeland and have returned there in increasing numbers, first in 1992 following the separation of Namibia from South Africa, and then in 1996 following the government slaughter of cattle after the outbreak of bovine pleuropneumonia.

Though staunchly traditional, the Herero participate actively in the developing economy of Botswana. They market cattle to the Botswana Meat Commission, send their children to school, and engage in wage employment. They herd cattle, goats and sheep, but it is the cattle that are important culturally and economically. In the double-descent system of Herero, they, historically, distinguish between the patrilineal (*otuzo*) and matrilineal (*omanda*) cattle, the latter constituting the wealth of the herd. Veneration of cattle through the customs of the *otuzo* are essential to the daily life of the rural Herero in this study, and beef, milk and milk products provide a regular source of protein. Even the fat skimmed from fresh milk, processed into clarified butter, may be set aside for times of scarcity. Bulk grains, especially ground corn, supplement this dietary base. Goats, chickens, garden produce, game and other wild foods are also consumed and bartered, but their economic importance is small compared with cattle except during drought. Thus, all the Herero we met desire cattle and measure their success in terms of cattle holdings.

The Ngamiland region is arid with seasonal rainfall averaging about 400 mm annually (Central Statistics Office, 1988). In 1988, population density ranged from 1 to 10 persons/km$^2$ in the most densely populated region of this study to 0.1 persons/km$^2$ in the more remote regions (Central Statistics Office, 1988). These areas are interspersed with large expanses of waterless and uninhabited regions. Drought years are not uncommon, and little or no rainfall for several years in a row is not unusual. The most recent severe drought prior to this study ended in 1987 and resulted in the decimation of livestock herds across Botswana. Disease can also take catastrophic tolls, and in 1996 the Herero described here lost their cattle to an outbreak of contagious bovine pleuropneumonia in this region previously free of it. Diffusion of the disease throughout Ngamiland provoked the slaughter of nearly 300 000 cattle there to prevent its spread to other parts of Botswana (M. Moorad, personal communication, November 1992). Given the cultural significance of cattle to the daily life of Herero and others there as well as its economic and dietary importance, the effect of this devastation on their health and life ways has no doubt been acute.

Rural Herero live in nominally patrilineal homesteads (*ozonganda,* sg. *onganda*) consisting of one or more small huts arranged around a cattle kraal. The smallest residential unit includes a man, his wife, and their small children. The largest homesteads incorporate 30 or 40 members that may include the additional wives of a man, the families of his junior brothers, and the families of sons. Each homestead is associated with a herd of cattle managed by the homestead's elder male. Elsewhere we have classified the homestead and the associated herd of cattle as the "male household" (Harpending and Pennington, 1990). Each wife within a homestead may form her own "female household" that is physically and somewhat economically autonomous from other female households within that homestead. She will have her own set of huts, use only milk from cattle allocated to her, and feed only children belonging to her female household.

Although homestead residence is patrilineal, in practice we have found that descent traced through matrilineal kin to be more pervasive. This is because large homesteads tend to split along maternal blood ties, large numbers of children are born outside of marriage who then "belong" to mother's father's homestead, and women often return to their natal homesteads after divorce and widowhood.

Most cattle in the homestead are owned by its residents, but many are not. At marriage, when women move to their husbands' homesteads, they leave their cattle behind, and their father or brothers will tend them. Children who have been fostered out may own cattle in both their father's homestead and their residential homestead. Rights in children born to unmarried women may be purchased, with cattle, by fathers and years may pass before these cattle, as well as cattle promised in bridewealth and other transactions, are transferred to the herds of their owners. We have also discovered that many cattle are associated with a homestead "on loan."

Homestead members, under the management of the elder male, tend all of these cattle – they water them, rear their calves, and drink their milk. However, unlike among other cattle pastoralists such as the Turkana (Gulliver, 1955), herding is hardly a labor-intensive activity. During the day cattle roam freely, untended, foraging for grass. At night calves and sometimes their mothers are kraaled near the homestead. Cows typically arrive here on their own at milking time to nurse their young.

Pools of rainwater provide most of needed water during the rainy season (roughly November through February) and a month or two afterward, but once these dry up, or if there is drought, livestock must be watered from wells. In two regions of our study water was lifted by hand from wells dug by hand, and this constitutes the hardest work of Ngamiland cattle ranching. Early in the dry season livestock visit the wells almost daily but as the grass gets eaten away near the homesteads and wells they must forage further and further away and seek water less frequently. In the third area of our study Herero watered livestock communally from a diesel-powered borehole. Following the rains in Angola the Okavango River fills and provides a preferred alternative water source for livestock from late July or August in this region.

### Data collection

#### Homestead censuses

We obtained reproductive histories of adult residents and visitors present during our visit to the three communities using methods described in Pennington and Harpending (1993). In brief, we asked subjects to provide the names, residence, and years of birth and death of all spouses, children, parents, fostered-in children, and siblings. Most born after 1960 knew at least their year of birth and many the exact day of birth of their children. The years of birth of others were obtained using Herero year names or from known age mates. We also censused homesteads to identify *de facto* members who were away socializing, at school or work, or on business. We asked hut owners (see above about female households) to name the members of her household, and all were asked to recall members currently away. We obtained reproductive histories of absent homestead members from other residents. We checked these data against information we collected in 1987–89 and 1990.

Because of the kin-based residential system, the reproductive histories ascertained nearly all homestead residents and the second-round census ascertained only a handful of additional members. While it is unlikely that we missed more than a few residents at the households, we were uncertain where many actually resided. Herero, however, rarely felt ambiguous about the homestead to which they belonged. Children whose parents were ever married to each other belonged to the homesteads of their fathers. About 40% of all children are born to unmarried women. A child in these circumstances "belongs" to mother's father unless mother later marries the child's father. The paternal rights in a small proportion of children are purchased, with cattle, by the childrens' fathers. These children then belong to their fathers' homesteads. About 40% of children are also fostered-out at some point in their childhod. Some fostered-out children are raised by relatives and by unrelated Herero for a few months or years, while others live indefinitely with the foster-parent. At maturity boys invariably returned to the homestead to which they belong. An adult male clearly belonged either to his own homestead or to the homestead of his father or elder brother, and I classified male residence accordingly. Female fostered-out children, however, typically remained in their foster parents' homesteads until marriage, often bearing several children prior to marriage that were also reared by the foster parent. We classified these fostered-out women as residents of the foster parents' homestead as long as they still resided there, reasoning that the foster parents' household was supporting the women and her children. Using similar reasoning, we classified currently married women as residents of their husbands' homestead and widowed or divorced women as residents of the homesteads in which they currently had an established household (i.e., in which they maintained huts). Although widowed women belonged to the homesteads of their husbands' families and divorced women belonged to their fathers' homesteads, many women returned to their natal homestead after the death of a spouse or continued to live in the homestead of the former husband. We classified children as residents of the same homestead as the parent or foster parent rearing them.

Using these criteria we were able to assign residence to individuals in a consistent, replicable and meaningful way. Although individuals living away from their homesteads for extended periods were counted as homestead dependents, their residential assignment corresponds to the primary resource pool, the homestead herd, to which they contributed and which they used. These homesteads averaged 13 members, ranging from 4 to 34.

*Herd size*

Herero support themselves economically in a number of ways. In this analysis we used the number of cows per homestead resident to classify individuals as either "cattle rich" or "cattle poor." Many, especially young men, engage in wage employment and sell goats, sheep and other small stock locally. We were unable to obtain much information about the amount of these resources or how they were used, especially since most were employed for only a few months at a time. All Herero desire cattle, however, and most convert other resources into cattle when possible. They measure their own wealth in terms of cattle, and we associated successful cattle ranchers with community prestige. Because cattle holdings constitute the only visible and measurable marker of household economic success in this community and has been correlated with household well-being throughout Botswana, it is an appropriate measure of Herero economic status.

We obtained information about the size of homestead herds in several ways. First, we attended milking sessions to identify individual cows (Herero name them). Later we interviewed families about how each of these cows, and any we missed, were acquired, their genealogy, parity, age, and owner as well as total herd size. We also asked for the number of cattle sold that year. We also visited individual wells and counted the number of cows, steers, bulls and calves that were watered. For some families we were more successful than with others at ascertaining herd size using these methods, and as the dry season progressed our success diminished due to changing grazing conditions. In one region of our study grass was scarce so cattle were not showing up daily for water, and those that did straggled in the entire day. Some families were no longer milking their cows or kraaling them.

In the region with the borehole, the cattle were communally watered. We counted the number of steers, bulls, cows and calves at the wells and received the counts of individual homestead holdings from the men running the borehole. The sum of the family holdings matched our own counts. Our estimates of herd sizes in this region are also probably smaller than actual holdings. Although grazing conditions were still fairly good here, we visited this community later in the dry season. Consequently, the proportion of cows still being milked had dwindled, and we ascertained fewer of them at the milking sessions. Meanwhile the nearby Okavango Delta was flooding and cattle were going there for water instead of the borehole.

Because of the differences in grazing conditions and the timing of our visit to each community our herd counts, which certainly underestimate the true sizes of Herero cattle holdings, are not directly comparable among the three regions. We expect the differences in grazing conditions in the three regions differentially affected marketing strategies as well as the proportions of actual holdings we observed in each region. Herero market most steer prior to the dry season when they are in peak condition and when they will bring the highest price from the Botswana Meat Commission. There also may be differences among families within regions with regard to marketing strategies and their candidness with us about how many they sold.

Nevertheless, the herd counts are suitable for ranking homesteads regionally. We assumed that all families were equally likely to omit cattle in the self-reports and that in each region an equal proportion of cattle were likely to be missed at the well. Finally, we assumed that omissions contribute noise rather than bias. We examined a number of ways of determining homestead wealth and settled on one measuring the number of cows per resident. The number of cows appearing at the kraals seemed similar to numbers observed at watering for herds with fewer than 30 or 40 cows (according to well counts) but never seemed to exceed this in herds with larger numbers of cows. Consequently cow counts from wells seemed more reliable. Using this criteria, most homesteads had about the same number of cows per capita (one or two), and we classified these as cattle poor while a smaller fraction clearly had many more (four to six) that we classified as cattle rich. That is, the assignment of wealth status occurred at the only natural break in the rankings. The per capita cattle counts implied by this method is somewhat less that the 12 per capita we estimated previously (Pennington and Harpending, 1993), probably because of the difference in residence classification and the more complete counts, provided to us by the Botswana veterinary department, used in that calculation.

In the three communities, I classified 18 homesteads as cattle poor and 10 as cattle rich. Age structure in the two groups is similar. In cattle-poor homesteads, 22% were born in 1988–92, 43% in 1973–87, 28% in 1933–72 and 7% before 1933. In cattle-rich homesteads the proportions were 23% in 1988–92, 41% in 1973–87, 33% 1933–72 and 3% before 1933.

Although we used the number of cows as a marker for wealth differences among families within regions, other classification criteria, such as the number of steer or total herd size per resident or per child produced similar rankings. Classification criteria based on the numbers of cows in herds seemed most appropriate for several reasons. First, cows are rarely sold so the number of cows owned by a family fluctuates much less year to year than the number of steers. Second, the cow counts are also more complete than counts of the total herd because we saw them at milking time and at the well and heard about them in the genealogies. Finally, number of cows reflects total herd productivity since they are the primary source of future herd growth. We included all cows managed by the homestead. We were unable to reliably determine individual use of resources resulting from individual cows.

### Anthropometrics

I measured the weight, stature, upper arm circumference and triceps skinfold of everyone present at the homesteads we censused as well as individuals at a number of neighboring homesteads and villages. We measured 828 Herero of various ages in Ngamiland, though we report here only on 117 Herero children born between 1982 and 1992 who lived in homesteads of known size and wealth. Eighty-three children lived in cattle-poor homesteads and 36 lived in cattle-rich homesteads. We had to visit each homestead several times to measure children away during earlier visits and visited villages to find children staying there during the school week.

We weighed children in light clothing without shoes using a bathroom scale set on a board leveled in the sand. Weights were recorded to the nearest kilogram. We weighed children too small to stand in their mothers' arms and subtracted mothers' weights from the total. We measured the height of children by asking them to stand erect on the board (there were no flat surfaces to stand against) and recorded their height to the nearest centimeter using a steel tape. We recorded the height of all children able to stand on their own or with assistance. Our protocol included the use of more precise measuring instruments but these failed prior to our arrival in the field. Our measurements are therefore subject to greater variance, but they are unbiased and therefore useful for intergroup comparisons.

Following standard techniques described in Frisancho (1990), we measured upper arm circumferences ($C$) using a steel tape and triceps skinfolds using Lange skinfold calipers. In brief I measured the upper arm circumference to the nearest 0.1 cm at the midpoint of the upper right arm. I measured the triceps skinfold thickness ($T$) in millimeters at this same point and report here on the average of three readings. Upper arm muscle area (UMA) was computed from the formula $(C-T\pi)^2/4\pi$ (Frisancho, 1990). We did not obtain these measurements on two toddlers who were too fearful of the skinfold calipers.

**Results**

Figures 8.1 through 8.5 compare triceps skinfolds, upper-arm muscle area, height and weight by age and weight for height by age and wealth. The solid lines in all figures are the results of smoothing the Herero data using loess regression in SPlus 2000 by MathSoft, Inc. This smoothing technique estimates values using least squares criteria (Cleveland and Devlin, 1988). Each smoothed point is a kind of average of neighboring data points. A line's overall smoothness depends on the percentage of data points (the span) with which each point is averaged; each point in the span is weighted according to its distance from the fitted point. I chose spans only wide enough to smooth sharp fluctuations between years and fit a quadratic rather than a line if it significantly improved fits between the smooth and the data using criteria described later. The circles show the scatter of the data around the curves and in all cases illustrate a good fit between the smoothed line and pattern in the data.

The dotted lines in Figures 8.1 through 8.4 represent the 5th, 50th and 95th percentiles for Black children aged 1 to 10 years included in the first and second National Health and Nutrition Examination Surveys (NHANES) as reported by Frisancho (1990). The dotted lines in Figure 8.5 are the percentiles for children aged 2 to 11 years for boys and 2 to 10 years for girls. In these figures I have plotted the averages of the sexes reported separately by Frisancho.

### Herero growth relative to US norms

Comparisons of Herero growth patterns with standard reference data give a perspective from which to assess their overall health. The plot of triceps skinfolds in Figure 8.1 indicates that Herero children less than age 5 years have more body fat than same-aged Blacks included in the NHANES survey, and mean Herero values exceed the NHANES 50th percentiles throughout early childhood. Among Herero, fat levels appear to peak around age 2 years and decline throughout early childhood. Mean triceps skinfolds decrease until about age 5 years in the NHANES sample and then begin increasing again (Frisancho, 1990). The peak in Herero triceps among 2-year-olds is not anomalous as it is also evident in the NHANES data when alternative smoothing techniques are used (Himes and Hoaglin, 1989).

The impression of nutritional adequacy from triceps skinfolds is some-what contradicted by the plot of UMA in Figure 8.2. UMA is a marker of

194    R.L. Pennington

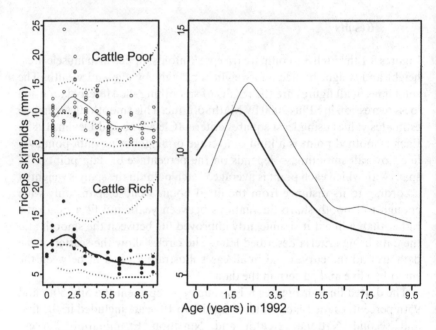

Figure 8.1. Herero triceps skinfolds by age and wealth. Solid lines: loess
smoothed Herero data (Splus 2000, MathSoft, Inc.). Dashed lines: 5th, 50th and
95th percentiles for Blacks, NHANES data (Frisancho, 1990). Circles: Herero
data. NHANES data are averages of boys and girls aged 1 to 10 years.

nutritional quality (protein adequacy) and quantity (caloric stores). Most
Herero children fall between the 50th and 5th percentiles. Several children
are well-below the 5th percentile, and mean values among the oldest
children are near the 5th percentile of the NHANES data. Similarly,
Figures 8.3 and 8.4 show that most Herero children are lighter and shorter
for age than those in the NHANES sample. Mean Herero values fall near,
and sometimes below, the 5th percentiles. That many Herero children may
be inadequately nourished is supported by the weight-for-height plot in
Figure 8.5. Mean Herero values lie between the 50th and 5th percentiles,
and a number of children fall below the 5th percentile.

The numbers of Herero at each age are too small to make meaningful
statistical comparisons with the NHANES data, but the figures make
several points. Herero children appear to grow slower and on a different
growth trajectory than children included in the NHANES survey. At a
given height, they are thinner yet have greater fat stores in the upper arms.
Low UMAs for age and low weights for heights, especially among the
oldest children, suggest that a number of Herero children receive inad-

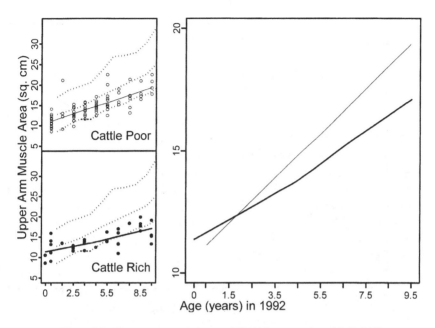

Figure 8.2. Herero upper muscle area (UMA) by age and wealth. Solid lines: loess smoothed Herero data (Splus 2000, MathSoft, Inc.). Dashed lines: 5th, 50th and 95th percentiles for Blacks, NHANES data (Frisancho, 1990). Circles: Herero data. NHANES data are averages of boys and girls aged 1 to 10 years.

equate nutrition. Many Herero women are obese (O'Keefe *et al.*, 1988) so it is clear that there are enough calories in this population, but the apparent growth lag of Herero children indicates that the distribution or quality of calories in childhood may be inadequate. Several children fall below the 5th percentiles for weight, height and UMA for age and for weight-for-height. To what extent these smaller and lighter children represent the natural low end of a distribution in this population is addressed later.

### *Herero health in regional perspective*

In a 1978 national health survey in Botswana, children aged 0 to 60 months whose weight for age fell below 80% of the mean of the Harvard standard were classified as "at-risk" (Kreysler, 1978). The percentages of at-risk children in the seven health regions of Botswana ranged from 23.1% to 27.8%, with Maun, the health region incorporating the Herero in this survey, reporting the lowest percentage. Among land use zones, cattle

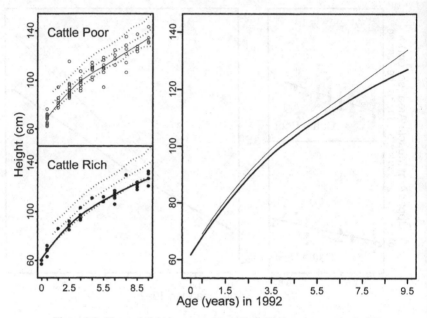

Figure 8.3. Herero height by age and wealth. Solid lines: loess smoothed Herero data (Splus 2000, MathSoft, Inc.). Dashed lines: 5th, 50th and 95th percentiles for Blacks, NHANES data (Frisancho, 1990). Circles: Herero data. NHANES data are averages of boys and girls aged 1 to 10 years.

posts in Botswana had 22.3% at risk, second only to high-income urban areas (12.5%). Within Ngamiland percent at-risk was lowest in Sehitwa (11.9%) and highest in Gomare (23.6%). Nationally, risk was highest among children aged 36 to 47 months and lowest among infants. Seasonally they were at their lowest levels in June (less than 23%) early in the dry season and peaked in November and December (nearly 30%) at the commencement of the rainy season.

Our methods differed from those used by the Botswana health ministry and so we can not report directly comparable figures. We also ascertained Herero children during the dry season only and at cattleposts (or at in nearby villages if temporarily away), so we expect them to have above-average nutritional status. We ascertained month and birth year of 47 of the 117 children already described as well as from 46 other Herero children living in Ngamiland. Seventeen percent of these Herero children age 5 or less ($n = 50$) weighed less than 80% of the Harvard standard, which falls within the ranges reported in Ngamiland. Twenty percent (10/50) of Herero children aged 24 months and less were most frequently low weight

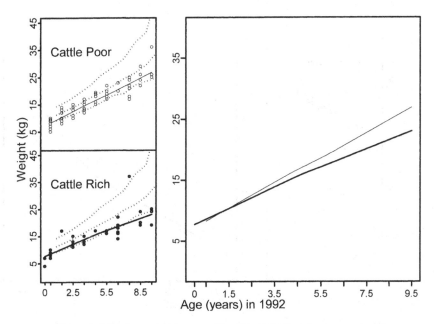

Figure 8.4. Herero weight by age and wealth. Solid lines: loess smoothed Herero data (Splus 2000, MathSoft, Inc.). Dashed lines: 5th, 50th and 95th percentiles for Blacks, NHANES data (Frisancho 1990). Circles: Herero data. NHANES data are averages of boys and girls aged 1 to 10 years.

for age; the frequency was less than 10% at the older ages but there were many fewer data. These findings indicate that our data are not regionally atypical. Like the NHANES data they also indicate that many children may be inadequately nourished using standard reference data.

Several studies have examined anthropometric indicators of health among !Kung Bushmen (Truswell and Hansen, 1976; Howell, 1979; Lee, 1979; Hausman and Wilmsen, 1985). The data from the earlier studies were recompiled by Hausman and Wilmsen (1985) who compared them with their own measurements. I compare these results with all Herero children whose birth month and year were both known.

Mean weights of !Kung girls and boys aged 1–4 years ranged from 9.2 to 12.5 kg, compared with 11 kg for Herero ($n = 60$). Mean heights of !Kung aged 2–4 years ranged from 82.9 to 94.6 cm, compared with 93 cm for Herero ($n = 32$). Among !Kung aged 5–9 years, mean weights ranged from 15.4 to 19.5 kg, compared with 18 kg among Herero ($n = 12$), and !Kung heights ranged from 108.7 to 113.1 kg, compared with 113 kg among Herero ($n = 12$).

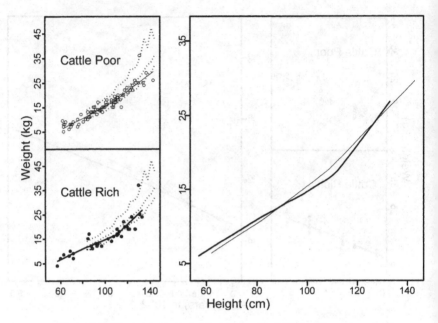

Figure 8.5. Herero weight by height and wealth. Solid lines: loess smoothed
Herero data (Splus 2000, MathSoft, Inc.). Dashed lines: 5th, 50th and 95th
percentiles for Blacks, NHANES data (Frisancho, 1990). Circles: Herero data.
NHANES data averages boys aged 2 to 11 years and girls 2 to 10 years. Only one
Herero child (female) less than age 2 years in the sample fell above the minimum
heights reported by Frisancho.

Triceps skinfolds of !Kung averaged 8.3 and 9.2 mm in boys and girls
aged 0 to 4 years and 6.8 and 7.4 mm in boys and girls aged 5 to 9 years
(Lee, 1979). In Herero the means were 10.0 mm in children aged 0 to 4
years ($n = 81$) and 8.8 mm in children aged 5 to 9 years ($n = 12$).

The mean values, of course, are affected by differences in age structure in
the age classes. In addition, adult stature of Herero exceeds the !Kung's by
more than 10 cm (O'Keefe et al., 1988). Nonetheless the comparisons
suggest that children in both groups are rather equally nourished. Herero
have greater fatfolds than !Kung (and the NHANES sample) but fall
within the ranges reported for !Kung in weight and height.

### Effect of wealth on child growth

The right-hand panels of Figures 8.1 to 8.5 compare the curves of cattle-
poor and cattle-rich households. These figures indicate that children from

Table 8.1. *Analysis of variance table for the effect of wealth on growth of children*[a]

| Model | ENP | RSS | *F* Value | Pr(F) |
|---|---|---|---|---|
| Triceps = age | 3.8 | 637 | 2.49 | 0.035 |
| Triceps = age × wealth | 7.7 | 569 | | |
| UMA = age | 2.3 | 526 | 2.97 | 0.040 |
| UMA = age × wealth | 4.7 | 488 | | |
| HT = age | 3.5 | 3735 | 1.71 | 0.154 |
| HT = age × wealth | 7.1 | 3507 | | |
| WT = age | 2.3 | 921 | 4.06 | 0.011 |
| WT = age × wealth | 4.7 | 834 | | |
| WT = HT | 3.0 | 566 | 1.45 | 0.225 |
| WT = HT × wealth | 6.1 | 537 | | |

[a] Using loess regression (Splus 2000 by MathSoft, Inc.), analysis compares probability that wealth explains portion of variance.
ENP, equivalent number of parameters; RSS, residual sum of squares; UMA, upper arm muscle area; HT, height; WT, weight.

cattle-poor homesteads grow better than children from cattle-rich homesteads. Cattle-poor children are somewhat taller and heavier than cattle-rich children at nearly all ages, with the largest differences occurring among the oldest children. Differences in weight-for-height by wealth are small, however, and follow no clear pattern. Triceps skinfolds indicate that cattle-rich infants have more body fat but that after 18 months cattle-poor children have more fat stores. The effect is also apparent in the plot of upper arm muscle area. Overall, the figures indicate that children living in less prosperous households grow better and are better fed, at least after age two, than children from wealthier households.

Although the numbers of children in this study are small, it is possible to assess statistically how well wealth explains differences between the groups using methods analogous to standard ANOVA techniques (Cleveland and Devlin, 1988). Given $y_i$, for $i = 1$ to $n$, observations on the dependent variables, and $x_i$, for $i = 1$ to $n$, observations of $p$ independent variables, the loess regression model is

$$y_i = g(x_i) + \gamma_i,$$

where $g(x)$ is a smooth function of $x$ and $\gamma_i$ is the error variable. Table 8.1 lists the residual sum of squares (RSS) and equivalent number of parameters (ENP) resulting from models with and without the independent variable *wealth*. ENP is a measure of the amount of smoothing. The

$F$-value is the test statistic resulting from the change in RSS between the two models.

Because there are few numbers of children at each age, it is unlikely that differences between cattle-rich and cattle-poor children at any two ages are meaningful. However, the $F$-test indicates that, across all ages, significant portions of the variation in triceps skinfolds, UMA and weight are explained by differences in cattle holdings among families. The contribution of wealth towards differences in height by age and weight by height could not be distinguished from stochastic variation. There are too few data to control for sex, but there are no differences between the sexes in any of the measurements in the absence of other variables (Pennington, 1997). There were also too few data to identify possible regional effects, but plots by region in which there were sufficient data showed the same tendency for cattle-poor children to surpass cattle-rich children in growth parameters.

Experimentation with parametric regression techniques produced the same answers. I report on the results of the loess local regression method because its only critical assumptions about the data are smoothness and an independent and normal distribution of error with constant variance. Graphical examinations of the residuals from the fit of these data to the models support the error distribution assumption. This method was also used to choose among smoothing parameters in the figures.

## Discussion

From the figures alone we reject the hypothesis that heterogeneity in infant and child mortality is attributable to low economic success, as measured by cattle holdings. It appears instead that poorer growth is associated with the most prosperous households. Triceps skinfolds and weights correspond to recent gains or shortfalls in calorie stores, and the results here indicate that poorer Herero children have more body fat and consequently were better fed than children living in more wealthy households at the time of our survey. Stature is the outcome of years of growth, and deficits in stature indicate nutritional deficits of the past. The trend among cattle-poor children towards greater stature at every age indicate they are indeed taller, but based on the loess regression results we cannot eliminate stochastic error as the cause of the difference. The regression results suggest that the apparent poorer nutritional status of cattle-rich children is too recent to have any affect or that the deficit is not nutritionally significant. It is also possible that measurement error in the field contributed to an excessively large stature variance and the lack of statistical significance.

The weight-for-height curve shows no sensible pattern. This statistic, however, combines two independent measures of growth and so is subject to two sources of measuring error. It nevertheless indicates that the shorter cattle-rich children have body weights similar to cattle-poor children when controlling for differences in height. Nonetheless, these children lag in height and weight compared with similarly aged children in the cattle-poor homesteads.

UMA is also derived from two body measurements (upper arm circumference and triceps skinfold thickness) and subject to multiple sources of measuring error but at a single site. The plot and regression results of UMA indicate that cattle-poor children have larger muscle masses and that the difference between the groups of children widens with age. UMA varies with energy and protein intake, and, like the differences in triceps skinfolds, these results suggest that children in cattle-poor homesteads were more adequately nourished than those living in more prosperous homesteads.

The consistent picture emerging from these data is that cattle-rich children are more poorly nourished than their cattle-poor age-mates. The figures indicate that the deficit begins during the second year, but data among younger children are sparse. The number of children living in cattle-poor homesteads is more than double the number living in cattle-rich homesteads, yet many more cattle-rich children fall below the 5th percentiles of the NHANES data in Figures 8.2–8.4 than in cattle-poor homesteads. The larger proportion of cattle-rich children in this category suggests that many more of them may be significantly nutritionally stressed. In any sufficiently large sample a proportion of healthy children should fall below the 5th percentile, but the disproportionate number of cattle-rich children in this category is indicative of meaningful differences in nutritional status between the two groups of Herero children.

The mortality rates of Herero children are highest during infancy and early childhood. The large deficits in growth apparent in the data occur most frequently among older cattle-rich children. These deficits should result in higher morbidity and mortality rates among older children in cattle-rich homesteads. Overall death rates are low among Herero children aged 5 to 10, and deaths are perhaps too rare to identify any risk factor during these ages. Because the growth curves cross over during early childhood, when mortality is relatively high, the effect of cattle holdings on nutritional status during ages 0 to 5 years is ambiguous. The cause of heterogeneity in infancy and early childhood remains unexplained.

Overall, caloric resources in this population must be plentiful because many adults are obese (O'Keefe *et al.*, 1988). The apparent differences in nutritional status between cattle-rich and cattle-poor children may reflect preferential distribution of resources within homesteads. Many Herero children are fostered-out to friends and relatives, and these children may receive fewer quality resources (Pennington and Harpending, 1993). We described above the homestead as the "male household" that contained smaller "female households" that were somewhat economically autonomous. Further research in the field is needed to assess economic differences among female households within homesteads.

Herero child growth patterns compare well with other cross-sectional samples locally and in Africa. In weight and height they resemble neighboring !Kung children, who also experience similar infant and early child mortality rates. They are similar in weight, height and adiposity to other African pastoralists such as the Datoga (Sellen, 1999) and Turkana (Little *et al.*, 1983) as well as much of sub-Saharan Africa (Cameron, 1991), but mortality during childhood is much higher in these groups. For example, mortality during infancy and childhood is 0.20 and 0.10 among the Datoga (Borgerhoff Mulder, 1992), several times higher than among recent cohorts of Herero.

## Conclusions

Herero children appear generally thin but healthy. Anthropometric assessment of their health indicates that they lag in height, weight and UMA for age and weight for height compared with Blacks included in the NHANES survey. Many fell below the 5th percentiles of the NHANES data, which may be indicative of severe nutritional stress. These children came disproportionately from cattle-rich homesteads. Triceps skinfolds in both groups of children are at or above the 50th percentiles at most ages. None were below the 5th percentile.

Weights and heights of Herero children less than 10 years old are comparable to those of neighboring !Kung. Survival of !Kung children increased to levels comparable to Herero children about the time !Kung began settling in large numbers among these and other cattle peoples. Previously I (Pennington and Harpending, 1988, 1993) attributed the better survival to improved weaning diets associated with sedentary living. The similarity in anthropometric measures of health as well as survival in the two groups further supports this conclusion.

Herero and !Kung child survival rates are low by African standards but high compared with Europeans. There appear to be sufficient calories available to Herero, yet the anthropometric indicators of health are similar to other African groups such as the Datoga with much higher levels of infant and child mortality. The large differences in mortality between Herero and other African pastoralists and between Herero and Europeans appear due primarily to care-independent causes of death, such as infectious disease. These findings do not dispute the synergistic role of nutritional adequacy on child survival already well established (Scrimshaw *et al.*, 1968) but predict that large reductions in child mortality will come in the future from regional improvements in environmental conditions rather than dietary changes.

The goal of the analysis was to examine the possibility that previously observed heterogeneity in infant and child mortality among Herero could be explained by socioeconomic differences among families. The nutritional status of children has been correlated with risk of morbidity and mortality in numerous settings. The hypothesis that children associated with more prosperous homesteads, as measured by cattle holdings, grow better is rejected. Children from homesteads with fewer cattle were significantly heavier and fatter and had larger UMAs. No significant differences in height and weight-for-height were evident. The findings reported here indicate that the children living in the most prosperous Herero homesteads are at greatest risk. These differences occur largely among older children who experience very low mortality risks. The effects of wealth on nutritional status among younger children are ambiguous. The cause of heterogeneity in infant and childhood mortality remains therefore unexplained. Further research needs to examine the allocation of resources within households.

### Acknowledgments

I thank all the Herero and Mbanderu for participating in this study and the Botswana government for permission to carry out research. I give special thanks to Muzeja Korujezu, Vinepeze Muinjo, Kaitira Ndjarakana, and Gakekgosi Otugile for their assistance in Botswana. I am grateful to Henry Harpending and Dennis O'Rourke for comments on this manuscript.

Field research was supported by National Science Foundation grants 9005813 and 9107587 to Henry Harpending and Jeffrey Kurland at The Pennsylvania State University. RLP received support while in the field from the National Institute of Child Health and Human Development through a grant for core support of population research (HD–05876) and a training grant (HD–07014) to the Center for Demography and Ecology, University of Wisconsin.

## References

Borgerhoff Mulder, M. (1992). Demography of pastoralists: preliminary data on the Datoga of Tanzania. *Human Ecology* 20, 383–405.

Cameron, N. (1991). Human growth, nutrition, and health status in sub-saharan Africa. *Yearbook of Physical Anthropology* 34, 211–250.

Central Statistics Office (1988). *Statistical Bulletin,* Vol. 13, No. 1. Gaborone, Botswana.

Cleveland, W.S. and Devlin, S.J. (1988). Locally weighted regression: an approach to regression analysis by local fitting. *Journal of the American Statistical Association* 83, 596–610.

Frisancho, A.R. (1990). *Anthropometric Standards for the Assessment of Growth and Nutrional Status.* Ann Arbor: University of Michigan Press.

Gulliver, P. 1955, *The Family Herds: A Study of Two Pastoral Tribes in East Africa, the Jie and Turkana.* London: Routledge and Kegan Paul.

Harpending, H. and Pennington, R. (1990). Herero households. *Human Ecology* 18, 417–439.

Harpending, H., Draper, P. and Pennington, R. (1990). Cultural evolution, parental care, and mortality. In *Health and Disease in Transitional Societies,* ed. A. Swedlund and G. Armelagos, pp. 241–255. South Hadley, MA: Bergin and Garvey.

Hausman, A.J. and Wilmsen, E.N. (1985). Economic change and secular trends in the growth of San children. *Human Biology* 57, 563–571.

Hill, A., Randall, S. and van den Eerenbeemt, M.-L. (1983). Infant and child mortality in rural Mali. Research Paper No. 83–5. London: Centre for Populations Studies.

Himes, J.H. and Hoaglin, D.C. (1989). Resistant cross-age smoothing of age-specific percentiles for growth reference data. *American Journal of Human Biology* 1, 165–173.

Howell, N. (1979). *Demography of the Dobe !Kung.* New York: Academic Press.

Kreysler, J. (1978). Report on National Nutritional Surveillance in Botswana: The System, Method and Results for 1978. Nutrition Unit, Ministry of Health, Republic of Botswana.

Lee, R. (1979). *The !Kung San.* Cambridge: Cambridge University Press.

Little, M.A., Galvin, K. and Mugambi, M. (1983). Cross-sectional growth of nomadic Turkana pastoralists. *Human Biology* 55, 811–830.

Litschauer, J. and Kelly, W. (1981). *The structure of traditional agriculture in Botswana.* Gaborone, Botswana: Division of Family Planning and Statistics.

O'Keefe, S.J.D., Rund, J.E., Marot, N.R., Symmonds, K.L. and Berger, G.M.B. (1988). Nutritional status, dietary intake and disease patterns in rural

Hereros, Kavangos and Bushmen in South West Africa/Namibia. *South African Medical Journal* **75**, 643–648.

Pennington, R. (1997). Bringing up children: fathers' contributions to successful child rearing among Kalahari Herero. An assessment based on juvenile growth patterns. Poster presented at the Annual Meeting of the American Association of Physical Anthropologists, St. Louis, April 1997.

Pennington, R. and Harpending, H. (1988). Fitness and fertility among Kalahari !Kung. *American Journal of Physical Anthropology* **77**, 303–319.

Pennington, R. and Harpending, H. (1993). *The Structure of an African Pastoralist Community: Demography, History, and Ecology of the Ngamiland Herero.* Research Monographs on Human Population Biology 11. Oxford: Clarendon Press.

Scrimshaw, N.S., Taylor, C.E. and Gordon, J.E. (1968). *Interactions of nutrition and infection.* World Health Organization Monograph No. 57. Geneva: WHO.

Sellen, D.W. (1999). Growth patterns among seminomadic pastoralists (Datoga) of Tanzania. *American Journal of Physical Anthropology* **109**, 187–209.

Trusswell, A.S. and Hansen, J.D.L. (1976). Medical research among the !Kung. In *Kalahari Hunter-Gatherers,* ed. R.B. Lee and I. DeVore, pp. 166–194. Cambridge, MA: Harvard University Press.

# 9 Ecology, health and lifestyle change among the Evenki herders of Siberia

WILLIAM R. LEONARD, VICTORIA A. GALLOWAY,
EVGUENI IVAKINE, LUDMILLA OSIPOVA AND
MARINA KAZAKOVTSEVA

## Introduction

Throughout the world, pastoral and herding subsistence strategies evolved and continue to persist in low-producing ecosystems where intensive agriculture cannot be supported. Research on the ecology of pastoral populations of Africa and the Andes has provided important insights into how these groups adapt to marginal environments characterized by limited and seasonally variable energy availability (e.g., Thomas, 1973; Galvin, 1992; Little and Leslie, 1999). To date, however, there has been relatively little research on the ecology and energetics of high-latitude (northern) herding groups. These groups are of particular interest because they are exposed to multiple ecological constraints that will influence energy balance and nutritional status. These stressors include: (1) low primary productivity, (2) severe cold stress, and (3) marked changes in day length. Limited primary productivity places constraints on food availability among northern pastoral groups. Conversely, cold stress and seasonal changes in photoperiod appear to increase daily energy requirements by elevating basal metabolic rates (see Rode and Shephard, 1995a; Shephard and Rode, 1996; Galloway et al., 2000).

With the shift away from a traditional lifestyle, there are important changes in both activity patterns and food availability that promote changes in health and nutritional status. Research among indigenous populations of the Canadian Arctic indicates that the transition from a subsistence lifestyle has resulted in dramatic declines in health and fitness levels (Rode and Shephard, 1994a,b; Shephard and Rode, 1996). This issue of how lifestyle change influences health has not been widely studied among indigenous populations of Siberia despite the fact that both ongoing and historic social and political changes in Russia have had a dramatic influence on the lifeways of these groups.

This chapter will examine aspects of ecology, energetics and health status of the Evenki reindeer herders of Central Siberia, one of the few remaining northern herding populations. We will first examine patterns of daily energy expenditure and dietary consumption among the Evenki. We will specifically consider how age, sex and residence location (i.e., those living in herding groups versus villages) all influence activity patterns, energy expenditure and dietary habits in the Evenki. Further, we will examine how the diet and energy expenditure of the Evenki compare with other indigenous herding and circumpolar populations. The implications of these patterns of expenditure and intake for body size, body composition, and plasma lipid levels will then be considered. We will compare anthropometric and lipid data for the Evenki with those of other indigenous circumpolar populations to see if they display similar patterns of chronic health risk. Together, these data provide important insights into the nutritional ecology of the Evenki, how they compare with other indigenous herding and circumpolar populations, and what the health consequences of historic and ongoing lifestyle changes have been.

### Background

The Evenki population is among the largest of the indigenous Siberian groups. Today, there are approximately 30 000 Evenki in Siberia, most of them (79.8%) continuing to live in isolated rural areas (Hannigan, 1991). Traditionally, the Evenki were socially organized into patrilineal named clans that served as herding units. During Stalinist times, they were forcibly reorganized into cooperative settlements and herder groups called brigades (Vasilevich, 1946; Uvachan, 1975; Forsyth, 1992). As a result of collectivization, reindeer herds were no longer held by families, but rather, were placed into communal herds and controlled by the cooperatives. Collectivization also resulted in the construction of permanent settlements with health and educational facilities. The construction of the collective villages, along with greater articulation with the rest of Siberia, resulted in a shift away from the traditional nomadic lifestyle, and the loss of central aspects of their indigenous culture (e.g., shamanism, language). The transition also resulted in a restructuring of men's and women's subsistence roles; that is a transition from the relatively unstructured division of labor (characteristic of most pastoral populations), to one in which male and female division of labor is much more well defined (with men now being largely responsible for herding the reindeer).

Today during the spring and summer months, each cooperative divides into 8–10 herding brigades that disperse over the *taiga*, living in temporary tepee structures known as *chums*. Figure 9.1(a,b) shows examples of life in the herding brigades. In the winter, most of the children and many of the women congregate in the central villages, and herding crews of mostly men and a few women rotate in and out of the herding brigades. A small proportion of the cooperative (e.g., the elderly, pregnant women and very young children) remain in the village year round (see Figure 9.2).

Our research among the Evenki between 1991 and 1995 has been conducted in the Stony Tunguska region of the Evenk Okrug in Central Siberia (see Figure 9.3). We have worked with Evenki from the cooperative settlements of Surinda and Poligus, and eight of their associated herding brigades. Additionally, we have worked with a small group of more urbanized Evenki living in the district capital of Baykit. At the time of our research, the population sizes for Surinda and Poligus were 613 and 481, respectively. According to government records, 95% of the Surinda population was indigenous (largely Evenki), as compared with 48% in Poligus. In contrast, Baykit's population was 5187, with only about 7% of the population (363 individuals) being Evenki.

### Energy expenditure and dietary consumption

Daily energy expenditure and dietary consumption were measured on 102 Evenki subjects (46 men; 56 women) ranging in age from 16 to 53 years. Energy expenditure was determined using daily heart-rate (HR) monitoring. This method involves first measuring HR and energy expenditure (EE) for individual subjects under resting and exercising conditions. These data are used to produce a HR versus EE relationship for each subject. After establishing the HR–EE relationship, subjects wear a HR monitor for an entire day. The monitor records and stores minute-by-minute HR values. The subject's HR data are downloaded to a computer and then converted to energetic equivalents based on HR versus EE relationship. An example of a daily HR trace is shown in Figure 9.4.

Dietary consumption was measured using standard 24-hour dietary recalls. Dietary information was obtained for the same day on which the HR data were collected. Subjects were asked about the types and amounts of foods consumed during the previous day. Energy and macronutrient intakes were then determined using food composition tables for Asia (FAO, 1972, 1982).

Figure 9.1(a) Evenki family riding through the Central Siberian *taiga* as they move their encampment during the late summer. (b) Evenki family standing in front of the cooking fire of brigade encampment.

Figure 9.2. Two Evenki mothers and their children in the village of Surinda.

### Sex and age differences

Table 9.1 presents information on body size, total energy expenditure (TEE; kcal/day) and total energy intake (TEI) for Evenki men and women. Despite their small body size, the Evenki have relatively high levels of energy expenditure. Evenk men spend an average of 2681 kcal/day, significantly more than the average of 2067 kcal/day in women ($p < 0.05$). In both sexes, energy expenditure is higher among individuals under the age of 30 years; however, these age differences are not statistically significant.

Using the World Health Organization's Physical Activity Level index (PAL = TEE/BMR (basal metabolic rate)) to assess levels of work intensity, we find that Evenki men under the age of 40 years have PALs of 1.75–1.81, commensurate with moderate–heavy activity levels (see FAO/WHO/UNU, 1985, p. 78; James and Schofield, 1990). Over the age of 40, work levels among men decline to the light–moderate level (PAL = 1.65). Among Evenki women, activity levels tend to be more variable with age. Activity levels are light–moderate among Evenki girls between 15 and 17 years of age (PAL = 1.57), moderate–heavy among women between 18 and 40 years (PALs = 1.70–1.75), and very low (near minimum maintenance levels) for women over 40 years (PAL = 1.38).

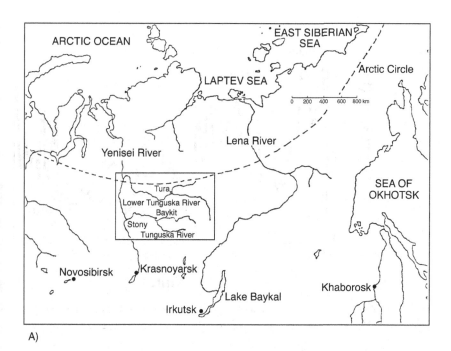

A)

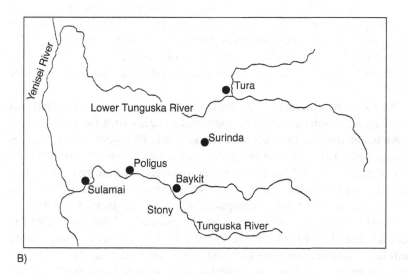

B)

Figure 9.3. Map showing the locations of (a) the Stony Tunguska region within Central Siberia, and (b) the study communities of Surinda, Poligus and Baykit.

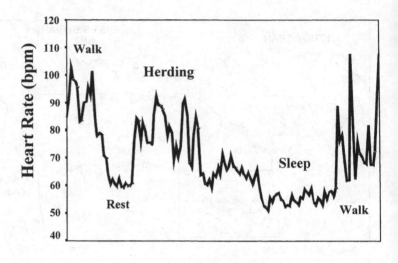

Figure 9.4. Twenty-four-hour heart rate trace for a typical day of an Evenki male.

Despite having relatively high activity levels, the Evenki consume sufficient energy to meet their needs during the summer months. Daily energy consumption averages 2962 kcal/day in men and 2676 kcal/day in women (n.s.; see Table 9.1). In both sexes, energy intake is greatest in the 18–29 years age cohort, and declines in the older cohorts. However, differences in energy intake between age groups are not statistically significant for either men or women.

As for energy balance (i.e., intake/expenditure), it appears that, on average, all groups are meeting their energy needs with the exception of boys under the age 18, who have intake levels 15% below their requirements. The largest positive energy balance is seen among women over 40, whose intakes are 49% above expenditure levels. Overall, Evenki women have a larger average energy surplus than men, consuming 37% more energy than they expend, as compared with +18% for men ($p < 0.05$).

Table 9.2 presents information on the protein and fat content of the Evenki diet. The Evenki derive about 30% of their daily energy from animal foods (reindeer meat and wild game; Leonard *et al.*, 1996). Consequently, protein intakes in the Evenki are quite high, averaging 121 g/day in men and 112 g/day in women (n.s.). Evenki men and women both consume an average of ~2.1 g of protein per kilogram of body weight, about 2.5 times the minimum requirements of the FAO/WHO/UNU

Table 9.1. *Daily energy expenditure and energy intake of Evenki males and females by age group*

| Sex and age (years) | n | Body size | | Energy expenditure | | | | Energy intake | |
|---|---|---|---|---|---|---|---|---|---|
| | | Stature (cm) | Weight (kg) | BMR (kcal/day) | TEE (kcal/day) | PAL (TEE/BMR) | | TEI (kcal/day) | Energy balance (TEI/TEE) |
| *Males* | | | | | | | | | |
| <18 | 3 | 161.2 ± 8.1 | 55.9 ± 5.9 | 1715 ± 233 | 3074 ± 1077 | 1.76 ± 0.39 | | 2480 ± 1859 | 0.85 ± 0.61 |
| 18–29 | 20 | 163.4 ± 5.3 | 59.3 ± 6.6 | 1639 ± 154 | 2840 ± 641 | 1.75 ± 0.42 | | 3280 ± 1187 | 1.20 ± 0.50 |
| 30–39 | 14 | 158.0 ± 4.4 | 59.4 ± 7.4 | 1410 ± 285 | 2480 ± 822 | 1.81 ± 0.66 | | 2958 ± 1291 | 1.29 ± 0.66 |
| 40 and older | 9 | 156.5 ± 5.3 | 55.8 ± 4.3 | 1558 ± 405 | 2511 ± 857 | 1.65 ± 0.58 | | 2421 ± 1282 | 1.09 ± 0.75 |
| Total | 46 | 160.3 ± 5.8** | 58.4 ± 6.4** | 1558 ± 277** | 2681 ± 768** | 1.75 ± 0.52 | | 2962 ± 1281 | 1.18 ± 0.60* |
| *Females* | | | | | | | | | |
| <18 | 7 | 154.3 ± 6.5 | 51.9 ± 6.0 | 1298 ± 108 | 2032 ± 408 | 1.57 ± 0.31 | | 2716 ± 1565 | 1.35 ± 0.80 |
| 18–29 | 23 | 150.2 ± 4.6 | 50.0 ± 5.0 | 1352 ± 219 | 2296 ± 721 | 1.74 ± 0.65 | | 2836 ± 994 | 1.35 ± 0.61 |
| 30–39 | 13 | 147.7 ± 4.3 | 50.2 ± 4.8 | 1124 ± 182 | 1897 ± 482 | 1.72 ± 0.52 | | 2399 ± 771 | 1.32 ± 0.30 |
| 40 and older | 13 | 148.6 ± 5.4 | 60.5 ± 9.2 | 1332 ± 212 | 1849 ± 486 | 1.38 ± 0.23 | | 2621 ± 1107 | 1.49 ± 0.72 |
| Total | 56 | 149.8 ± 5.2 | 52.7 ± 7.5 | 1288 ± 214 | 2067 ± 607 | 1.63 ± 0.52 | | 2676 ± 1046 | 1.37 ± 0.62 |

Sex differences are significant at: *$p < 0.05$; **$p < 0.01$.

All values are mean ± SD. BMR, basal metabolic rate; TEE, total energy expenditure (kcal/day); PAL, physical activity level (TEE/BMR); TEI, total energy intake (kcal/day).

214    *W.R. Leonard* et al.

Table 9.2. *Dietary protein and fat intakes of Evenki males and females by age group*

| Sex (age) | n | Protein (g/day) | Fat (g/day) | Protein[a](%) | Fat[a] (%) |
|---|---|---|---|---|---|
| *Males* | | | | | |
| < 18 | 3 | 90.0 ± 86.4 | 35.1 ± 22.4 | 13.0 ± 4.0 | 13.7 ± 4.9 |
| 18–29 | 20 | 126.4 ± 57.3 | 82.5 ± 53.3 | 15.2 ± 4.2 | 22.3 ± 9.6 |
| 30–39 | 14 | 120.0 ± 52.7 | 64.3 ± 49.5 | 16.9 ± 5.0 | 20.8 ± 12.6 |
| 40 and older | 9 | 119.1 ± 71.7 | 53.0 ± 41.4 | 19.3 ± 3.0 | 18.7 ± 7.3 |
| Total | 46 | 120.6 ± 59.4 | 68.2 ± 49.2 | 16.4 ± 4.5 | 20.6 ± 9.9 |
| *Females* | | | | | |
| < 18 | 7 | 105.5 ± 5.9 | 67.9 ± 58.9 | 14.8 ± 3.9 | 19.9 ± 8.2 |
| 18–29 | 23 | 125.2 ± 58.9 | 68.2 ± 32.7 | 17.8 ± 5.2 | 21.4 ± 7.0 |
| 30–39 | 13 | 89.6 ± 43.0 | 59.6 ± 32.8 | 14.8 ± 4.9 | 21.5 ± 7.1 |
| 40 and older | 13 | 112.5 ± 49.5 | 81.0 ± 38.1 | 18.0 ± 6.3 | 28.1 ± 4.7 |
| Total | 56 | 111.9 ± 55.9 | 69.1 ± 36.6 | 16.8 ± 5.4 | 22.7 ± 7.2 |

All values are Mean ± SD.
[a] % Protein and % Fat; percent of dietary energy derived from protein and fat, respectively.

(1985) recommendations (0.8 g/kg). Dietary protein contributes an average of 16–17% of dietary energy in both men and women. Among men, there appears to be an age-related increase in the contribution of protein to the diet, with the percent of energy derived from protein increasing from 13% in the 16–18 years cohort to over 19% for men over 40 years ($p = 0.07$). In contrast, women do not show significant age differences in protein consumption.

Despite having a high-meat diet, fat consumption in the Evenki is quite modest, averaging 68–69 g/day in both sexes. Fat contributes 21% of dietary energy in men, and about 23% in women. These low dietary fat intakes reflect the fact that both the reindeer meat and wild game being consumed by the Evenki are quite lean. Women show significant age-related increases in the percent of energy derived from fat ($p < 0.03$), whereas men do not.

### Differences associated with residence location

Lifestyle factors associated with residence location also appear to have an important influence on diet and energy expenditure among the Evenki. Table 9.3 shows patterns of daily energy expenditure and dietary intake by residence location for Evenki men and women. Among men, those living in

Table 9.3. *Influence of residence location on energy expenditure
and dietary intake among Evenki men and women* [a]

| Residence Location | n | TEE (kcal/day) | TEI (kcal/day) | Energy balance (TEI/TEE) |
|---|---|---|---|---|
| *Males* | | | | |
| Brigades | 12 | 2693 ± 227 | 3863 ± 357 | 1.43 ± 0.18 |
| Surinda | 18 | 2886 ± 175 | 2968 ± 277 | 1.13 ± 0.14 |
| Poligus | 13 | 2496 ± 220 | 2215 ± 360 | 1.04 ± 0.18 |
| Baykit | 3 | 2210 ± 436 | 2318 ± 687 | 1.06 ± 0.35 |
| *p*-value | | 0.35 | 0.02 | 0.45 |
| *Females* | | | | |
| Brigades | 2 | 1985 ± 426 | 2706 ± 772 | 1.35 ± 0.45 |
| Surinda | 23 | 2124 ± 122 | 2954 ± 226 | 1.43 ± 0.13 |
| Poligus | 26 | 2123 ± 119 | 2404 ± 219 | 1.25 ± 0.13 |
| Baykit | 5 | 1545 ± 294 | 2795 ± 534 | 1.75 ± 0.28 |
| *p*-value | | 0.33 | 0.39 | 0.29 |

[a] Values have been adjusted for differences in age and body weight through analysis of covariance. All values are Mean ± SE.

the herding brigades and in the more traditional cooperative village (Surinda) have the highest expenditure levels, both averaging between 2700 and 2900 kcal/day. Daily energy expenditure in the less traditional and more ethnically diverse settlements (Poligus and Baykit) is lower, averaging between 2200 and 2500 kcal/day. Residence location appears to have less of an influence on energy expenditure among Evenki women. Those living in the brigades, Surinda, and Poligus have, similar expenditure levels, with averages ranging between 2000 and 2200 kcal/day. Women living in Baykit are more sedentary, with TEEs averaging less than 1600 kcal/day.

Residence location also influences dietary patterns. Surprisingly, among men we find much *higher* energy intakes in the more traditional settlements. Daily energy consumption averages ~3900 kcal/day in the brigades and ~3000 kcal/day in Surinda, as compared with 2200–2300 kcal/day in Poligus and Baykit. In contrast, the percent of dietary energy derived from fat and protein are highest in Baykit, the most urbanized community (fat = 28%; protein = 20%).

Among the women, there is less systematic variation in energy consumption. Women of the brigades, Surinda, and Baykit all have energy intakes between 2700 and 2950 kcal/day, whereas those of Poligus consume ~2400 kcal/day. Protein consumption shows considerable variation among the different residence groups, being lowest in the brigades (12% of energy)

Table 9.4. *Comparison of total energy expenditure and physical activity levels among selected circumpolar and pastoral populations.*

| Population | Sex | Body Weight (kg) | TEE (kcal/day) | PAL | Reference |
|---|---|---|---|---|---|
| Evenki (Siberian Herders) | M | 58.4 | 2681 | 1.8 | Present study |
| | F | 52.7 | 2067 | 1.6 | |
| Igloolik Inuit (Arctic hunters) | M | 65.0 | 3010 | 1.8 | Godin and Shephard |
| | F | 55.0 | 2350 | 1.8 | (1973) |
| Turkana (African pastoralists) | M | 65.4 | 2160 | 1.3 | Galvin (1985) |
| | F | 52.4 | 1733 | 1.3 | |
| Quechua (Andean agropastoralists) | M | 56.2 | 2012 | 1.3 | Leonard (1991) |
| | F | 48.9 | 1611 | 1.3 | |
| Aymara (Andean agropastoralists) | M | 54.8 | 2655 | 2.0 | Kashiwazaki *et al.* |
| | F | 50.6 | 2280 | 1.9 | (1995) |

and highest in Surinda (20%). This pattern suggests that within the herding brigades there may be differential distribution of meat to the men.

The sex differences in intake and expenditure patterns among the different settlements combine to produce marked differences in energy balance (see Table 9.3). Among the men, the largest positive energy balance is seen among the brigade dwellers, whose intakes exceed their expenditures by >40%. Those living in Surinda consume 10–15% more energy than they consume, whereas those of Poligus and Baykit have intakes that match their expenditure. In contrast, the pattern is almost the reverse among Evenki women. Energy consumption exceeds expenditure by 25–35% in the brigades and in Poligus, 43% in Surinda, and 75% in Baykit.

### Interpopulational comparisons

#### Energy expenditure and activity levels

Table 9.4 compares body size and daily energy expenditure of the Evenki to those of other Arctic (Inuit) and herding (Turkana, Quechua, Aymara) populations. Average TEE and PAL for the Evenki are greater than those of the Quechua and Turkana, and lower than those of the Inuit and

Aymara. Sex differences in energy expenditure among the Evenki are high in comparison with these other populations. Sexual dimorphism in body weight is relatively low in the Evenki, and yet TEE is more than 600 kcal/ day greater in men than in women. In contrast, Turkana men and women differ by an average of 13 kg, but show only a ~430 kcal difference in energy expenditure. The differences in TEE between Inuit men and women are similar to those seen in the Evenki ($\Delta = 660$ kcal); however, the Inuit also show much greater sex differences in body weight ($\Delta = 10$ kg).

The marked sex difference in daily energy expenditure and activity seen in the Evenki appears to be reflective of the division of labor associated with the collectivization. With their traditional, clan-based herding system, the Evenki were nomadic, moving with their encampments seasonally with the herds. Similar to other nomadic herding populations (e.g., Turkana and Quechua), the division of labor in traditional Evenki society was relatively fluid (see Service, 1963; Fondahl, 1998). In addition to their domestic roles (e.g., food preparation, child care), Evenki women were active participants in herding and preparation of the reindeer skins. Men were responsible for herding and breeding the reindeer, as well as hunting and trapping.

Collectivization resulted in a major restructuring of activities, having a most profound impact on women's roles (Fondahl, 1995). Women were encouraged to remain in the central villages, spending little or no time in the herding brigades (Fondahl, 1998). Moreover, even when in the brigades, the women were primarily responsible for maintenance of the encampment, being largely excluded from herding activities. These patterns are evident in the data presented in Table 9.3. In the brigade sample there are many more men than women ($n = 12$ males versus 2 females), and the sex difference in energy expenditure is quite high ($\Delta > 700$ kcal).

*Dietary consumption*

Table 9.5 compares the diet of the Evenki to those of other circumpolar groups. The Evenki have the highest daily energy intakes of the seven groups presented in Table 9.5. Evenki men consume 2962 kcal/day, as compared with the average of 2360 kcal (range: 1541–2754 kcal) for the other groups. The differences are similar in females, as Evenki women consume an average of ~700 kcal/day more than women from other circumpolar groups (2676 kcal for Evenki versus 1980 kcal for other groups).

Protein intakes of the Evenki fall in the middle of the range for circumpolar populations: higher than those of the Greenland Inuit (from Bang

Table 9.5. *Comparison of dietary energy, protein and fat intakes among selected circumpolar populations*

| Population | Sex | Energy (kcal/day) | Protein (g/day) | Fat (g/day) | Protein (%) | Fat (%) | Reference |
|---|---|---|---|---|---|---|---|
| Evenki (Siberia) | M | 2962 | 120.6 | 68.2 | 16.4 | 20.6 | Present study |
| | F | 2676 | 111.9 | 69.1 | 16.8 | 22.7 | |
| Inuit (Alaska) | M/F | 1990 | 148.0 | 110.8 | 23.8 | 39.8 | Draper (1977) |
| Inuit (Greenland) | M | 1541 | 84.4 | 69.6 | 23.0 | 39.0 | Bang *et al.* (1980) |
| Inuit (Baffin Island) | F | 2179 | 136.0 | 86.0 | 26.2 | 35.5 | Kuhnlein (1991) |
| Alaskan Natives | M | 2754 | 127.0 | 117.0 | 19.4 | 38.2 | Nobmann *et al.* (1992) |
| | F | 1945 | 90.0 | 81.0 | 19.4 | 37.5 | |
| Chukchi (Siberia) | M | 1949 | 163.0 | 93.0 | 26.0 | 32.0 | Nobmann *et al.* (1991) |
| Sámi (Finland) | M | 2462 | 123.1 | 106.7 | 20.0 | 39.0 | Näyhä *et al.* (2002) |
| | F | 1816 | 90.8 | 78.7 | 20.0 | 39.0 | Haglin (1991) |

*et al.*, 1980) and lower than those of the Alaskan Inuit (Draper, 1977) and the Chukchi of Siberia (Nobmann *et al.*, 1991). However, protein comprises a smaller portion of the diet in the Evenki, contributing 16–17% of dietary energy, as compared with 19–26% in the other populations.

The Evenki also derive a smaller percent of their energy from fat, ~20–25% as compared with 30–40% in the other groups. Thus, the Evenki diet is distinct from those of most other circumpolar populations in having a relatively larger carbohydrate component (~60% of energy). This difference partly reflects the fact that most of the groups in Table 9.5 live in more northern (e.g., tundra) ecosystems than the Evenki. Additionally, the collective system has, until recently, provided the Evenki villages and herding brigades with access to non-local foods such as sugar, flour, rice, and buckwheat. Thus, the greater carbohydrate consumption of the Evenki relative to the Inuit, Chucki and Sámi is a product of greater access to both locally available plant foods and market goods.

**Variation in nutritional status: body size and composition**

To evaluate variation in body size and composition standard anthropometric dimensions were taken on a sample of 375 adults (172 men, 203 women) ranging in age from 18 to 79 years. Measurements included stature (cm), weight (kg), upper arm circumference (cm), and four skinfold measurements (triceps, biceps, subscapular, and suprailiac). Two derived indices were calculated: (1) the body mass index (BMI) and upper arm muscle area (UMA) (see Frisancho, 1990). Additionally, percent body fatness was estimated from the sum of the four skinfolds using the age- and sex-specific equations presented in Durnin and Womersley (1974).

*Age and sex differences*

Gender differences in activity patterns and energy expenditure are reflected in differences in relative body size and adiposity between Evenki men and women. Table 9.6 presents mean anthropometric dimensions for Evenki men and women by age group. Evenki men and women are quite diminutive in stature (158 and 148 cm, respectively), both falling below the 3rd percentile relative to US normative data (e.g., Frisancho, 1990; Leonard *et al.*, 1994*b*). For body weight, on the other hand, women average ~53 kg, and approximate the 15–20th percentile relative to US norms, while men,

Table 9.6. *Anthropometric characteristics of Evenki men and women by age group*

| Sex and age (years) | n | Stature (cm) | Weight (kg) | BMI (kg/m$^2$) | UMA(cm$^2$) | Sumskin (mm) | Body fat[a] (%) |
|---|---|---|---|---|---|---|---|
| *Males* | | | | | | | |
| 18–29 | 62 | 161.2±7.1 | 57.1±6.6 | 22.0±2.3 | 38.2±7.7 | 27.4±8.7 | 11.2±3.4 |
| 30–39 | 50 | 157.6±5.4 | 57.6±7.7 | 23.1±2.5 | 40.0±7.0 | 33.4±12.5 | 16.7±3.6 |
| 40–49 | 30 | 155.4±5.1 | 54.6±5.7 | 22.6±1.8 | 37.7±7.7 | 32.1±10.2 | 17.9±4.3 |
| 50–59 | 15 | 155.0±7.9 | 51.9±6.9 | 21.6±2.8 | 31.9±6.1 | 25.4±11.0 | 15.1±5.6 |
| 60 and older | 15 | 150.7±4.6 | 52.4±6.4 | 23.1±3.1 | 28.7±5.4 | 35.9±25.1 | 18.8±8.8 |
| Total | 172 | 157.7±6.9 | 55.9±7.0 | 22.5±2.5 | 37.3±7.9 | 30.5±12.7 | 15.0±5.3 |
| *Females* | | | | | | | |
| 18–29 | 87 | 149.4±6.1 | 49.4±7.4 | 22.1±3.4 | 26.3±5.9 | 58.5±23.5 | 27.9±5.6 |
| 30–39 | 52 | 146.6±4.5 | 51.4±8.1 | 23.9±3.9 | 30.2±7.1 | 70.6±24.9 | 31.9±4.7 |
| 40–49 | 39 | 147.5±5.6 | 57.5±11.9 | 26.4±5.3 | 32.3±9.6 | 87.8±32.9 | 37.0±5.6 |
| 50–59 | 18 | 145.8±7.1 | 59.5±13.9 | 27.9±5.4 | 34.5±8.2 | 87.5±34.3 | 39.6±6.3 |
| 60 and older | 7 | 146.2±5.3 | 65.0±22.6 | 30.0±7.7 | 36.5±13.2 | 99.6±45.6 | 41.6±5.1 |
| Total | 203 | 147.9±5.8 | 52.9±10.8 | 24.2±4.8 | 29.5±8.1 | 71.0±30.4 | 32.1±7.0 |

All values are mean ± SD. BMI, body mass index; UMA, upper arm muscle area.
[a] Percent body fatness determined from sum of skinfolds based in the age- and sex-specific equations of Durnin and Womersley (1974).

averaging ~56 kg, fall at about the 5th US percentile. Evenki women show significant age-related increases in body weight, with women under the age of 30 years averaging below 50 kg, whereas those over the age of 60 years average 65 kg. In contrast, men show modest declines in weight, averaging 57–58 kg in the 18–29 and 30–39 year cohorts, and ~52 kg in the 50–59 and 60 and older cohorts.

Looking at the body mass index (BMI = weight (kg)/height (m)$^2$), we find that Evenki women are relatively heavier than Evenki men, and show substantially higher risks of obesity. According to recent NIH (1998) guidelines, the optimal BMI range (i.e., no increased risk of cardiovascular disease) for adult men and women is 18.5–24.9, with BMIs of 25.0–29.9 being classified as "overweight", and BMIs of 30 and above being classified as obese. Among the Evenki, 34.5% of women have BMIs $\geqslant$ 25, as compared with 9.9% of men ($p < 0.001$). Similarly, 11.3% of Evenki women are obese based on their BMI, versus only 1.7% of men ($p < 0.001$). Evenki men and women also show different age-related changes in BMI. Among men, the average BMIs in all of the age cohorts fall within the optimal range. Women, on the other hand, show significant age-related increases in BMI ($p < 0.001$), with mean values for women 40 years and older falling above 25. Thus, as shown in Figure 9.5, dramatic sex differences in BMI are evident over the age of 40.

Variation in body fatness mirrors that of the BMI. Women have significantly higher levels of body fatness than men, and show greater age-related increases in fatness. These patterns are evident from both the summed skinfold measures as well as the estimated percent body fat levels (see Table 9.6, Figure 9.6).

### Differences associated with residence location

Anthropometric differences for Evenki men and women by residence location are presented in Table 9.7. The patterns of variation are consistent with the variation observed in energy intakes and expenditure. Among men, body weights and BMIs are lower in the more urbanized villages (i.e., Poligus and Baykit), differences that are consistent with the lower levels of energy intake and dietary adequacy observed in these villages (see Table 9.3). Conversely, among women, body weight and body fatness are greatest in Baykit. This pattern is also consistent with the energetics data in that Baykit women had the lowest levels of energy expenditure and highest ratio of energy intake to energy expenditure.

Thus, the lifestyle differences associated with residence location appear

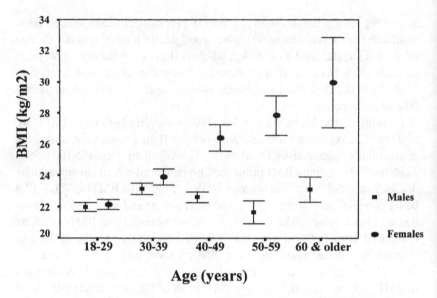

Figure 9.5. Age differences in the body mass index (BMI; mean ± SE) for Evenki men and women. Evenki women show significant age-related increases in BMI (*p* <0.001), whereas men do not. Mean BMIs for women over the age of 40 years fall in the overweight or obese range.

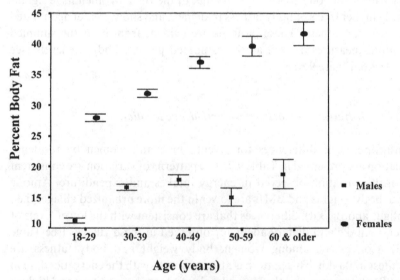

Figure 9.6. Age differences in percent body fatness (mean ± SE) for Evenki men and women. Both sexes show significant age-related increases in body fatness; however, the increases are much greater in women than in men.

Table 9.7. *Influence of residence location on the anthropometric characteristics of Evenki men and women*[a]

| Residence | n | Stature (cm) | Weight (kg) | BMI (kg/m$^2$) | UMA (cm$^2$) | Sumskin (mm) | Body Fat (%) |
|---|---|---|---|---|---|---|---|
| *Males* | | | | | | | |
| Brigades | 80 | 157.2 ± 0.7 | 56.5 ± 0.8 | 22.9 ± 0.3 | 37.3 ± 0.9 | 28.4 ± 1.4 | 13.9 ± 0.5 |
| Surinda | 51 | 158.5 ± 0.9 | 56.8 ± 1.0 | 22.6 ± 0.3 | 37.8 ± 1.1 | 34.1 ± 1.7 | 16.2 ± 0.7 |
| Poligus | 35 | 157.2 ± 1.0 | 54.2 ± 1.1 | 21.9 ± 0.4 | 37.4 ± 1.3 | 30.3 ± 2.1 | 15.7 ± 0.8 |
| Baykit | 6 | 160.6 ± 2.5 | 50.8 ± 2.8 | 19.6 ± 0.9 | 32.3 ± 3.1 | 28.3 ± 5.1 | 13.8 ± 1.9 |
| *p*-value | | 0.40 | 0.08 | 0.007 | 0.42 | 0.09 | 0.04 |
| *Females* | | | | | | | |
| Brigades | 37 | 145.9 ± 0.9 | 53.7 ± 1.7 | 25.2 ± 0.7 | 31.4 ± 1.3 | 72.5 ± 5.0 | 32.4 ± 1.0 |
| Surinda | 77 | 148.6 ± 0.6 | 53.7 ± 1.1 | 24.3 ± 0.5 | 29.1 ± 0.9 | 76.1 ± 3.2 | 33.0 ± 0.6 |
| Poligus | 62 | 146.9 ± 0.7 | 50.8 ± 1.3 | 23.6 ± 0.5 | 29.9 ± 1.0 | 60.9 ± 3.5 | 30.2 ± 0.7 |
| Baykit | 26 | 150.7 ± 1.1 | 54.2 ± 2.0 | 23.7 ± 0.8 | 27.0 ± 1.5 | 79.9 ± 5.4 | 34.1 ± 1.0 |
| *p*-value | | 0.002 | 0.28 | 0.33 | 0.16 | 0.004 | 0.003 |

[a] Values have been adjusted for differences in age through analysis of covariance. All values are mean ± SE.

Table 9.8. *Comparison of body size and body composition of adult males and females from three circumpolar populations: the Evenki (Central Siberia), Inuit (Northwest Territory, Canada) and Nganasan (Northern Siberia)*

| Measure | Age group | Males | | | Females | | |
|---|---|---|---|---|---|---|---|
| | | Evenki | Inuit | Nganasan | Evenki | Inuit | Nganasan |
| Stature (cm) | 20–29 | 160.7 | 164.3 | 164.0 | 149.0 | 153.2 | 153.0 |
| | 30–39 | 157.6 | 163.7 | 166.0 | 146.6 | 153.8 | 150.0 |
| | 40–49 | 155.4 | 162.7 | 164.0 | 147.5 | 152.8 | 153.0 |
| | 50–59 | 155.0 | 163.6 | – | 145.8 | 150.8 | – |
| | 60 and older | 150.7 | 159.7 | – | 146.2 | 147.5 | – |
| Weight (kg) | 20–29 | 57.6 | 65.0 | 61.0 | 50.4 | 54.6 | 61.6 |
| | 30–39 | 57.6 | 65.1 | 61.2 | 51.4 | 59.1 | 63.9 |
| | 40–49 | 54.6 | 74.4 | 64.6 | 57.5 | 67.6 | 69.3 |
| | 50–59 | 51.9 | 72.5 | – | 59.5 | 72.2 | – |
| | 60 and older | 52.4 | 69.6 | – | 65.0 | 61.6 | – |
| BMI (kg/m²) | 20–29 | 22.3 | 24.1 | 22.7 | 22.7 | 23.3 | 26.1 |
| | 30–39 | 23.1 | 24.2 | 23.6 | 23.9 | 25.0 | 28.4 |
| | 40–49 | 22.6 | 28.2 | 23.3 | 26.4 | 28.9 | 29.6 |
| | 50–59 | 21.6 | 27.1 | – | 27.9 | 31.7 | – |
| | 60 and older | 23.1 | 27.3 | – | 30.0 | 28.3 | – |
| % Body fat | 20–29 | 11.4 | 13.4 | 9.6 | 28.6 | 26.5 | 30.0 |
| | 30–39 | 16.7 | 16.9 | 13.0 | 31.9 | 32.9 | 35.1 |
| | 40–49 | 17.9 | 23.4 | 13.5 | 37.0 | 38.3 | 37.5 |
| | 50–59 | 15.1 | 23.3 | – | 39.6 | 41.7 | – |
| | 60 and older | 18.8 | 20.5 | – | 41.6 | 37.4 | – |

Inuit data are from Rode and Shephard (1994*a*); Ngansan data are from Rode and Shephard (1995*c*).

to be having divergent effects on the nutritional status of Evenki men and women. Among men, the more urbanized, less traditional lifestyle is associated with lower levels of dietary adequacy which contribute to lower body weights. For women, on the other hand, the more urbanized lifestyle is associated with a larger surplus of dietary energy and higher levels of body fatness.

### *Interpopulational comparisons*

Table 9.8 compares adult body size and composition in three circumpolar populations: the Evenki, the Igloolik Inuit of the Northwest Territories of Canada (from Rode and Shephard, 1994*a*) and the Nganasan of the

Taymir Peninsula of Northern Siberia (from Rode and Shephard, 1995*c*). The Evenki are significantly smaller than either the Inuit or Nganasan. Evenki men are, on average, 3 to 9 cm shorter and 3 to 20 kg lighter than their age peers from the other two populations. The pattern is similar for females. Evenki women are between 1 and 6 cm shorter and 4 to 12 kg lighter than their Inuit and Nganasan age peers.

In contrast, the Evenki are more similar to the Nganasan and Inuit in terms of BMI and percent body fatness. Evenki men have almost identical BMIs (22–23 kg/m$^2$) and slightly higher body fat levels than their Nganasan peers (11–19% versus 10–14%). Relative to Inuit men, the Evenki are lighter (BMIs = 22–23 versus 24–28 kg/m$^2$) and leaner (fatness = 11–19% versus 13–23%). Among women, BMIs tend to be lower in the Evenki, except in the oldest age cohort (60 years and older). In terms of body fatness, all three groups show similar levels and age-related increases, with fatness levels ranging from 27% to 30% in the youngest women, to 40% and above among women over the age of 50 years.

These comparisons indicate that the Evenki are relatively small in comparison with other indigenous circumpolar populations. Some of this reflects the fact that the Evenki have remained relatively more isolated than other northern populations over the past century. Indeed, our previous work has shown that there is little or no evidence of a secular trend in stature among the Evenki over the past century (see Leonard *et al.*, 1996). Thus, unlike the majority of traditional human societies, the Evenki have not shown increased growth rates in recent decades.

In terms of BMI, and body fatness, the Evenki are more similar to the Inuit and Nganasan. Moreover, it appears that the Nganasan and Evenki show similar patterns of sex differences in BMI and body fatness. That is, in both Siberian groups, women show considerably higher risks of obesity than men, whereas, among the Inuit, such sex differences are less evident. These comparisons indicate that greater risks of obesity among women may be a pattern that is evident among indigenous Siberian populations in general.

### Plasma lipid levels

Plasma lipid levels (total, high-density lipoprotein (HDL) and low-density lipoprotein (LDL) cholesterol, and triglycerides) were measured on a sample of 208 Evenki adults (100 men, 108 women) between 18 and 76 years of age. Blood samples were collected by venipuncture and separated by centrifugation into their constituent parts. Plasma samples were analyzed by the Clinical Laboratories of the University of Kansas Medical

Table 9.9. *Plasma lipid levels of Evenki males and females by age group*

| Sex and age (years) | n | Tchol (mmol/l) | HDL-C (mmol/l) | LDL-C (mmol/l) | Triglycerides (mmol/l) |
|---|---|---|---|---|---|
| *Males* | | | | | |
| 18–29 | 39 | 3.46 ± 0.68 | 1.23 ± 0.39 | 1.84 ± 0.74 | 0.97 ± 0.35 |
| 30–39 | 27 | 3.81 ± 0.97 | 1.19 ± 0.35 | 2.26 ± 0.82 | 0.93 ± 0.47 |
| 40–49 | 17 | 3.83 ± 0.84 | 0.95 ± 0.29 | 2.64 ± 0.76 | 0.86 ± 0.35 |
| 50–59 | 9 | 3.98 ± 0.74 | 1.31 ± 0.33 | 2.17 ± 0.67 | 1.11 ± 0.40 |
| 60 and older | 8 | 3.66 ± 0.50 | 0.97 ± 0.30 | 2.37 ± 0.58 | 0.79 ± 0.23 |
| Total | 100 | 3.68 ± 0.80* | 1.16 ± 0.37 | 2.16 ± 0.79** | 0.93 ± 0.38 |
| *Females* | | | | | |
| 18–29 | 49 | 3.69 ± 0.86 | 1.15 ± 0.33 | 2.28 ± 0.84 | 0.81 ± 0.45 |
| 30–39 | 22 | 3.96 ± 1.03 | 1.10 ± 0.41 | 2.60 ± 0.87 | 1.03 ± 0.65 |
| 40–49 | 22 | 4.08 ± 1.32 | 1.09 ± 0.34 | 2.74 ± 1.11 | 1.03 ± 0.84 |
| 50–59 | 13 | 4.29 ± 0.86 | 1.07 ± 0.30 | 2.97 ± 0.72 | 0.93 ± 0.34 |
| 60 and older | 2 | 5.07 | 1.25 | 3.69 | 0.65 |
| Total | 108 | 3.92 ± 1.021 | 1.12 ± 0.34 | 2.55 ± 0.92 | 0.91 ± 0.58 |

Sex differences are significant at: $*p < 0.05$; $**p < 0.01$.
All values are Mean ± SD. Tchol, total cholesterol; HDL-C, high-density lipoprotein cholesterol; LDL-C, low-density lipoprotein cholesterol.

Center (M. Chiga, Director). Cholesterol and triglyceride levels were measured enzymatically using the Kodak EKTACHEM system (Allain *et al.*, 1974). HDL cholesterol was measured after precipitation of other lipoprotein fractions using dextran sulfate and magnesium (Finley *et al.*, 1978). LDL cholesterol levels were calculated using the Friedwald equation (Friedwald *et al.*, 1972).

### Age and sex differences

Age and sex differences in plasma lipid levels are presented in Table 9.9. Despite consuming a diet rich in meat and other animal products, plasma lipid levels of the Evenki are low relative to US norms. Recent data from the US National Center for Health Statistics (NCHS, 1999) indicate that average total cholesterol levels of adult men and women range from 4.7 to 4.9 mmol/l for men and women in their 20s, to 5.7–6.0 mmol/l for adults over 55 years.

Total cholesterol levels of Evenki women are significantly higher than those of Evenki men (3.92 mmol/l versus 3.68 mmol/l $p=0.02$). Evenki women also have significantly higher LDL cholesterol levels (2.55 mmol/l versus 2.16 mmol/l; $p < 0.001$) These differences are consistent with the

Table 9.10. *Correlates of plasma lipid levels among Evenki men and women*

| Independent variables | Plasma lipid levels | | | |
|---|---|---|---|---|
| | Tchol | HDL-C | LDL-C | Triglycerides |
| *Males* | | | | |
| Body weight | 0.27** | 0.16 | 0.11 | 0.32** |
| BMI | 0.24** | 0.10 | 0.14 | 0.20* |
| Sum of four skinfolds | 0.19* | −0.02 | 0.18 | 0.16 |
| Age | 0.20* | −0.14 | 0.27** | −0.02 |
| Residence location[a] | 0.23* | −0.17* | 0.43** | −0.09 |
| *Females* | | | | |
| Body weight | 0.17* | −0.08 | 0.21* | 0.08 |
| BMI | 0.25** | −0.08 | 0.29** | 0.19* |
| Sum of four skinfolds | 0.17* | −0.13 | 0.25** | 0.08 |
| Age | 0.28** | −0.01 | 0.33** | 0.10 |
| Residence location[a] | 0.10 | −0.11 | 0.19* | 0.21* |

[a] Coded: 1, Brigades; 2, Surinda; 3, Poligus; 4, Baykit.
Pair-wise correlations are significant at: $*p < 0.05$; $**p < 0.01$.

higher dietary adequacy and greater risk of obesity observed among the Evenki women. Moreover, this pattern of higher plasma lipid levels among women is the opposite of the pattern observed in the United States and other populations in which women tend to have lower cholesterol levels (see NCHS, 1999).

Both men and women show significant age-related increases in total and LDL cholesterol levels. HDL cholesterol levels show modest declines with age among men. Such reductions are likely attributable to declining activity levels.

### Correlates of plasma lipid levels

The correlates of plasma lipid levels in Evenki men and women are presented in Table 9.10. The measures of body mass and composition (i.e., weight, BMI and sum of skinfolds) are all significantly correlated with total cholesterol levels in both men and women ($r = 0.17$–$0.27$). All three of these measures are also significant predictors of LDL cholesterol levels in women, but not in men. Triglyceride levels are significantly correlated with BMI in women and with both BMI and weight in men. However, none of the body composition measures are significantly correlated with HDL cholesterol level in either sex.

Table 9.11. *Multiple regression analyses of the correlates of total cholesterol and LDL cholesterol among Evenki males*

|  | Dependent variables | |
| --- | --- | --- |
| Independent variables | Tchol beta weights | LDL-C beta weights |
| Age | 0.28** | 0.36*** |
| Body weight | 0.32*** | 0.19* |
| Residence location | 0.28** | 0.48*** |
| Model $R^2$ | 0.21*** | 0.32*** |

Significant at: * $p < 0.05$; *** $p < 0.01$; *** $p < 0.001$.

Age and residence location are also important correlates of plasma lipid levels. Both sexes show significant age-related increases in total and LDL cholesterol levels ($r = 0.20-0.33$). HDL and triglyceride levels, on the other hand, do not vary significantly with age. Additionally, a more urbanized lifestyle also appears to influence lipid levels in both sexes. Among men, those living in Poligus and Baykit have significantly higher total and LDL cholesterol levels and significantly lower HDL levels than those living in Surinda and the herding brigades. Among women, those of the more urbanized settlements have higher LDL and trigylceride levels. These patterns likely reflect differences in activity patterns and diet composition among the different residence locations (see Table 9.3). For both men and women, those living in Baykit have the lowest levels of energy expenditure. Baykit men have the diets with the highest percent fat content, whereas Baykit women consume the largest surplus of energy above their daily requirements.

The joint influence of these different factors on lipid levels was explored using stepwise multiple regression analyses. Table 9.11 presents the multiple regression models for explaining variation in total and LDL cholesterol level among Evenki men. In both models, age, body weight and residence location are significant predictors of lipid levels. Overall, the independent variables explain 21% of the variation in total cholesterol levels and 32% of the variation in LDL levels. For the women, no multivariate model was significantly better than the bivariate relationships.

### Interpopulational comparisons

Figure 9.7(a,b) compare total cholesterol levels of the Evenki men and women with those of the Inuit and Nganasan (from Rode and Shephard, 1995c; Rode *et al.*, 1995; Shephard and Rode, 1996). There is considerable

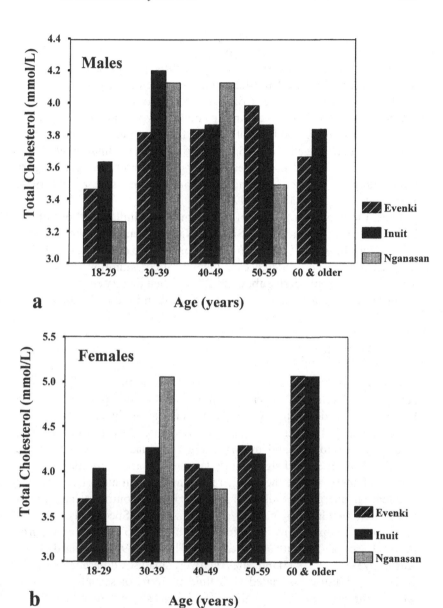

Figure 9.7. Comparison of mean total cholesterol levels by age group for (a) males and (b) females of three indigenous circumpolar populations (Evenki, Inuit, and Nganasan). The cholesterol levels of all three groups are similar, and are low relative to US values.

overlap in the cholesterol values, with none of the three groups being consistently the highest or lowest. In both sexes, the total cholesterol levels are low relative to US Norms. Men in all three groups have cholesterol levels ranging between 3.2 and 4.2 mmol/l. Female values tend to be higher in all three groups, and range from 3.4 to 5.1 mmol/l.

Cholesterol levels, however, are not uniformly low among indigenous circumpolar groups. The Inuit of Alaska, for example, have total cholesterol levels that range between 5.1 and 6.6 mmol/l (Feldman et al., 1975; Draper, 1980). The Sámi reindeer herders of Finland have even higher levels, ranging between 6.6 and 7.7 mmol/l (Bjorkstene et al., 1975; Westlund, 1981; Näyhä et al., 1994, 2002).

Thus, the cholesterol levels of the Evenki are similar to those of the Igoolik Inuit of Canada and the Siberian Ngansan, but lower than those of the Alaskan Inuit and Sámi of Finland. Some of the differences in cholesterol level may reflect differences in dietary fat consumption. As noted in Table 9.4, the Evenki derive about 20–22% of their dietary energy from fat, as compared with almost 40% among the Alaskan Inuit and Sámi.

#### Discussion

This chapter has examined aspects of energy expenditure and health status among the Evenki of Siberia, a northern herding population. The Evenki have moderate to high levels of energy expenditure, but consume sufficient energy to meet their daily requirements. Marked sex differences in energy expenditure and dietary energy adequacy are evident in the Evenki. Women expend significantly less energy, and consume a greater surplus of energy above their daily requirements than men. The large sex difference in energy expenditure among the Evenki contrasts sharply with the pattern seen in many herding/pastoral societies. Specifically, research on pastoral populations of Africa (Galvin, 1985, 1992) and agro-pastoral groups of South America (e.g., Thomas, 1973; Leonard, 1991; Kashiwazaki et al., 1995) indicates relatively small sex differences in daily energy expenditure associated with fluid patterns of sexual division of labor. Ethnographic data on the Evenki suggests that their traditional lifeway (prior to Soviet collectivization) was more similar to other pastoral groups; they were nomadic, with women participating heavily in the herding activities (see Service, 1963; Fondahl, 1998). However, after collectivization, men's and women's roles were more clearly divided, with men being almost exclusively responsible for herding activities (and thus being disproportionately represented in the herding brigades), and women being primarily responsible for domestic activities. These differences are clearly evident among the Evenki today, and contribute to a more sedentary lifestyle for Evenki women.

These lifestyle differences between Evenki men and women have important consequences for differences in risk of obesity and other chronic health problems. Evenki women are 3.5 times more likely to be overweight than Evenki men. Moreover, Evenki women show dramatic age-related increases in body weight and fatness. This pattern of women being at greater risk for obesity is one that is shared with other indigenous circumpolar populations (e.g., Inuit of Canada, Nganasan of Northern Siberia). However, the magnitude of the gender differences documented in the Evenki and Nganasan suggests that the unique historical influence of Soviet collectivization has and continues to have a profound impact on women's health in these groups. Additionally, body weights and levels of obesity of the Evenki are lower than those observed among indigenous circumpolar populations of North America (e.g., Inuit; Young and Sevenhuysen, 1989; Rode and Shephard, 1994a, 1995c; Young et al., 1995; Young, 1996). This difference likely reflects the fact that indigenous Siberians groups have remained more isolated than their North American counterparts over the past 40–50 years.

Evenki women also show poorer plasma lipid profiles than men, having significantly higher total and LDL cholesterol levels. These differences are consistent with the sex differences in energy expenditure, dietary adequacy, and body composition. Moreover, variation in age, body composition and lifestyle factors (i.e., residence location) all have significant influences on lipid levels in both men and women.

The marked differences in activity patterns, risk of obesity and lipid levels in the Evenki suggest that women may be at greater risk for cardiovascular diseases than men. An examination of mortality records from the Baykit district from 1982–94 supports this hypothesis. During this period, cardiovascular diseases were the leading cause of death among Evenki women, accounting for 33% of deaths compared with 19% of deaths among men. In contrast, accidents and violence were the leading cause of death among men, accounting for 56% of male deaths in comparison with 31% of female deaths.

Overall, this work underscores how historic and ongoing lifestyle changes in Siberia have influenced activity patterns, nutrition and health status in the Evenki herders. Changes associated with collectivization resulted in a restructuring of the social and economic roles of men and women. It appears these changes have had a greater impact on the lifestyle, biology and health of Evenki women. Relative to men, Evenki women have lower activity levels, are relatively heavier and fatter, have higher cholesterol levels, and are more likely to die of cardiovascular disease. These differences seem broadly to parallel those observed among other indigenous circumpolar groups.

Future health trends in the Evenki remain unclear, particularly in light of the current socioeconomic changes in Russia and, more specifically, within Siberia. Since the fall of the Soviet Union, the collective structure is being replaced by the traditional extended family ownership and control of the reindeer herds. Additionally, the villages and herding brigades have become increasingly isolated, because of the great reduction in plane and helicopter flights into Central Siberia. Hence the transport of food and medical supplies, and well as travel in the region, has been greatly restricted. The changes appear to be promoting the re-emergence of many traditional elements of the Evenki lifeway. Further research is needed to explore how these ongoing changes will influence the biology and health of the Evenki.

## Acknowledgments

We are most grateful to all the subjects who participated in our research. Comments by Dr Marcia Robertson substantially improved this chapter. This research was supported by grants from the Natural Sciences and Engineering Research Council of Canada (OGP–0116785), the US National Science Foundation (BSR–99101571), and the National Geographic Society.

## References

Allain, C.C., Poon, L.S., Chan, C.S.G., Richmond, W. and Paul, C.F. (1974). Enzymatic determination of total serum cholesterol. *Clinical Chemistry* **20**, 470–475.

Bang, H.O., Dyerberg, J. and Sinclair, H.M. (1980). The composition of Eskimo food in northwestern Greenland. *American Journal of Clinical Nutrition* **33**, 2657–2661.

Bjorkstene, F., Aromaa, A., Eriksson, A.W., Maatela, J., Kirjarinta, M., Fellman, J. and Tamminen, M. (1975). Serum cholesterol and triglyceride concentrations of Finns and Finnish Lapps: interpopulational comparisons and the occurrence of hyperlipidemia. *Acta Medica Scandinavica* **198**, 23–33.

Draper, H.H. (1977). The aboriginal Eskimo diet in modern perspective. *American Anthropology* **79**, 309–316.

Draper, H.H. (1980). Nutrition. In *The Human Biology of Circumpolar Populations,* ed. F.A. Milan, pp. 257–284. Cambridge: Cambridge University Press.

Durnin, J.V.G.A. and Womersley, J. (1974). Body fat assessed from total body density and its estimation from skinfold thicknesses: measurements on 481 men and women aged from 16 to 72 years. *British Journal of Nutrition* **32**, 77–97.

FAO (Food and Agriculture Organization) (1972). *Food composition table for use in East Asia.* Rome: FAO.

FAO (Food and Agriculture Organization) (1982). *Food composition table for use in the Near East.* Rome: FAO.

FAO/WHO/UNU (Food and Agriculture Organization/World Health Organization/United Nations University) (1985). *Energy and Protein Requirements.* WHO Technical Report Series No.724. Geneva: WHO.

Feldman, S.A., Ho, K.-J., Lewis, L.A., Mikkelson, B. and Taylor, C.B. (1975). Lipid and cholesterol metabolism in the Alaskan Arctic Eskimos. *Archives of Pathology* **94**, 42–58.

Finley, P.R., Schifman, R.B., Willams, R.J. and Lichti, D.A. (1978). Cholesterol in high-density lipoprotein: use of $Mg^{2+}$/dextran sulfate in its enzymatic measurement. *Clinical Chemistry* **24**, 931–933.

Fondahl, G. (1995). Legacies of territorial reorganization for indigenous land claims in northern Russia. *Polar Geography and Geology* **19**, 1–21.

Fondahl, G.A. (1998). *Gaining Ground? Evenkis, Land and Reform in Southeastern Siberia.* Needham Height, MA: Allyn & Bacon.

Forsyth, J. (1992). *A History of the Peoples of Siberia.* Cambridge: Cambridge University Press.

Friedwald, W.J., Levy, R.I. and Fredrickson, D.K. (1972). Estimation of the concentration of low-density lipoprotein cholesterol in plasma, without use of the preparative ultracentrifuge. *Clinical Chemistry* **18**, 499–502.

Frisancho, A.R. (1990). *Anthropometric Standards for the Assessment of Growth and Nutritional Status.* Ann Arbor, MI: University of Michigan Press.

Galloway, V.A., Leonard, W.R. and Ivakine, E. (2000). Basal metabolic adaptation of the Evenki reindeer herders of Central Siberia. *American Journal of Human Biology* **12**, 75–87.

Galvin, K.A. (1985). Food procurement, diet, activities and nutrition of Ngisonyoka Turkana pastoralists in an ecological and social context. PhD dissertation. State University of New York, Binghamton.

Galvin, K.A. (1992). Nutritional ecology of pastoralists in dry tropical Africa. *American Journal of Human Biology* **4**, 209–221.

Godin, G. and Shephard, R.J. (1973). Activity patterns of the Canadian Eskimo. In *Polar Human Biology,* ed. O.G. Edholm and E.K.E. Gunderson, pp. 193–215. Chichester: Heinemann.

Haglin, L. (1991). Nutrient intake among Saami people today compared with an old traditional Saami diet. In *Circumpolar Health 90,* ed. B.D. Post *et al.,* pp. 741–746. Winnipeg: Canadian Society for Circumpolar Health.

Hannigan, J. (1991). Statistics on the economic and cultural development of the northern aboriginal people of the USSR (for the period of 1980–1989). Ottawa: Bureau of Indian and Northern Affairs.

James, W.P.T. and Schofield, E.C. (1990). *Human Energy Requirements: A Manual for Planners and Nutritionists.* Oxford: Oxford University Press.

Kashiwazaki, H., Dejima, Y., Orias-Rivera, J. and Coward, W.A. (1995). Energy expenditure determined by the doubly labeled water method in Bolivian Aymara living in a high altitude agropastoral community. *American Journal of Clinical Nutrition* 62, 901–910.

Kuhnlein, H.V. (1991). Nutrition of the Inuit: A brief overview. In *Circumpolar Health 90*, ed. B.D. Post *et al.*, pp. 728–730, Winnipeg: Canadian Society for Circumpolar Health.

Leonard, W.R. (1991). Age and sex differences in the impact of seasonal energy stress among Andean agriculturalists. *Human Ecology* 19, 351–368.

Leonard, W.R., Comuzzie, A.G., Crawford, M.H. and Sukernik, R.I. (1994a). Correlates of low serum lipid levels among the Evenki herders of Siberia. *American Journal of Human Biology* 6, 329–338.

Leonard, W.R., Katzmarzyk, P.T., Comuzzie, A.G., Crawford, M.H. and Sukernik, R.I. (1994b). Growth and nutritional status of the Evenki reindeer herders of Central Siberia. *American Journal of Human Biology* 6, 339–350.

Leonard, W.R., Katzmarzyk, P.T. and Crawford, M.H. (1996). Energetics and population ecology of Siberian herders. *American Journal of Human Biology* 8, 275–289.

Little, M.A. and Leslie, P.W., ed. (1999). *Turkana Herders of the Dry Savanna.* Oxford: Oxford University Press.

NCHS (National Center for Health Statistics) (1999). *Health United States, 1999 with Health and Aging Chartbook.* Hyattsville, MD: National Center for Health Statistics.

NIH (National Institutes of Health) (1998). *Clinical Guidelines on the Identification, Evaluation, and Treatment of Overweight and Obesity in Adults.* NIH Publication no. 98-4083. Bethesda, MD: National Heart, Lung, and Blood Institute.

Näyhä, S., Sikkilä, K. and Hassi, J. (1994). Cardiovascular risk factor patterns and their association with diet in Saami and Finnish reindeer herders. *Arctic Medical Research* 53, 301–304.

Näyhä, S., Luoma, P., Lehtinen, S., Lehtimäki, T., Mosher, M.J. and Leppäluoto, J. (2002). Disease patterns in Sámi and Finnish populations: An update. In *Human Biology of Pastoral Populations*, ed. W.R. Leonard and M.H. Crawford, pp. 236–250, Cambridge: Cambridge University Press.

Nobmann, E.D., Mamieeva, F.R. and Rodigina, T.A. (1991). A preliminary comparison of the nutrient intakes of the Siberian Chukotka and Alaska natives. In *Circumpolar Health 90*, ed. B.D. Post *et al.*, pp. 752–755. Winnipeg: Canadian Society for Circumpolar Health.

Nobmann, E.D., Byers, T., Lanier, A.P., Hankin, J.H. and Jackson, M.Y. (1992). The diet of Alaska native adults: 1987–1988. *American Journal of Clinical Nutrition* 55, 1024–1032.

Rode, A. and Shephard, R.J. (1994a). Physiological consequences of acculturation: a 20-year study of fitness in an Inuit community. *European Journal of Applied Physiology* **69**, 516–524.

Rode, A. and Shephard, R.J. (1994*b*). Growth and fitness of Canadian Inuit: Secular trends 1970–1990. *American Journal of Human Biology* **6**, 525–541.

Rode, A. and Shephard, R.J. (1995*a*). A comparison of physical fitness between Igloolik Inuit and Volochanka nGanasan. *American Journal of Human Biology* **7**, 623–630.

Rode, A. and Shephard, R.J. (1995*b*). Basal metabolic rate of Inuit. *American Journal of Human Biology* **7**, 723–729.

Rode, A. and Shephard, R.J. (1995*c*). Modernization of lifestyle, body fat content and body fat distribution: a comparison of Igloolik Inuit and Volochanka nGanasan. *International Journal of Obesity* **19**, 709–716.

Rode, A., Shephard, R.J., Vloshinsky, P.E. and Kuksis, A. (1995). Plasma fatty acid profiles of Canadian Inuit and Siberian Ganasan. *Arctic Medical Research* **54**, 10–20.

Service, E.R. (1963). *Profiles in Ethnography*. New York: Harper & Row.

Shephard, R.J. and Rode, A. (1996). *The Health Consequences of "Modernization": Evidence from Circumpolar Peoples.* Cambridge: Cambridge University Press.

Thomas, R.B. (1973). *Human Adaptation to a High Andean Energy Flow System.* Occasional papers in anthropology, no. 7. University Park, PA: Pennsylvania State University.

Uvachan, V.N. (1975). *The Peoples of the North and their Road to Socialism.* Moscow: Progress Publishers.

Vasilevich, G.M. (1946). Drevneyshiye etnonimy Azii nazvaniya evenkiyskikh rodov. Sovetskaya Etnografiya, No. 4.

Westlund, K. (1981). Circumpolar non-infectious epidemiology. *Nordic Countries and Arctic Medicine Research Reports* **31**, 215–220.

Young, T.K. (1996). Obesity, central fat patterning, and their metabolic correlates among the Inuit of the Central Canadian Arctic. *Human Biology* **68**, 245–263.

Young, T.K. and Sevenhuysen, G. (1989). Obesity in Northern Canadian Indians: patterns, determinants, and consequences *American Journal of Clinical Nutrition* **49**, 786–793.

Young, T.K., Nikitin, Y.P., Shubnikov, E.V., Astakhova, T.I., Moffatt, M.E.K. and O'Neil, J.D. (1995). Plasma lipids in two indigenous arctic populations with low risk for cardiovascular diseases. *American Journal of Human Biology* **7**, 223–236.

# 10 Disease patterns in Sámi and Finnish populations: an update

SIMO NÄYHÄ, PAULI LUOMA, SAARA LEHTINEN, TERHO
LEHTIMÄKI, MARY JANE MOSHER AND JUHANI
LEPPÄLUOTO

## The Sámi of Finland and their health: previous investigations

The Sámi (formerly known as Lapps) are the indigeneous people of
Scandinavia and northern Russia. The number of the Sámi today is
estimated to be 50 000, of whom 4000 live in Finland (Beach, 1988). The
origin of the Sámi is unknown. They inhabited the inland regions of
Finland early in the last millenium and were gradually pushed northwards
with colonization, mixing with the Finns at the same time. The Sámi
population today is almost entirely concentrated in the far north, officially
termed the "Sámi district", which consists of the three northernmost
communities of Finland (Figure 10.1). Even here, the Finns have con-
stituted the main population over the last century, with the exception of
Utsjoki, which has remained predominantly Sámi. Up to the present day,
the principal occupations of the Sámi have been reindeer herding, fishing
and hunting but now they have been largely assimilated and mainly work
in sales, services and industry. Reindeer herding has also been practiced by
the Finns since the eighteenth century, and today approximately 85% of
the herders are Finns. Finland is the only country where most reindeer
herders are from the main population. In Sweden and northern Norway
reindeer herding is only allowed to the Sámi.

Due to the new legislation which came into force in the 1980s, the
seasonal migration of the herding families following the reindeer flocks has
ended with most herders now living in fixed dwellings. Although the
herding itself has been greatly mechanized, the work remains physically
strenous. Many phases, e.g., sorting the reindeer and marking the calves,
are still performed manually. Major health hazards now include snow-
mobile noise and vibration, cold, heavy lifting and accidents. To review the
health hazards of reindeer herding work and to arrange an appropriate
health care model for the herders, The Oulu Regional Institute of Occupa-

236

NORWAY

Figure 10.1. The study area. The Sámi district shaded.

tional Health surveyed all reindeer herders in Finland in 1986–89. The number of herders was 3720. Geographically, the survey covered the official reindeer herding area, i.e., the territory north of the Kiiminki river. Of the examinees 15% were Sámi and 85% were Finns.

Previously, the Sámi were known for an infant mortality 2–3 times higher than in the main population (Lewin et al., 1971), high occurrence of tuberculosis (Palva and Finell, 1976), rickets, echinococcosis, congenital dislocation of the hip, spondylolysis and osteoarthritis (Lewin and Hedegård, 1971) and almost no dental caries (Lewin et al., 1971). As with indigenous people elsewhere (Young, 1988), the Sámi have been undergoing an epidemiological transition, marked by declines in infant mortality, tuberculosis and other infectious diseases, but an increase in dental caries. Information on other "modern" diseases such as cancer and cardiovascular conditions is limited, but they probably constitute a significant proportion of morbidity among the Sámi nowadays. New hazards include environmental pollution, the health effects of which are poorly known in this population. This chapter summarizes some major findings from the

Finnish reindeer herders' survey and other relevant studies, concentrating on living habits, cardiovascular risk factors and potential effects of environmental pollution. Other health aspects of the Sámi population have been reported from the WHO Biological Programme performed in the late 1960s (Lewin and Eriksson, 1970).

## Living habits

### Smoking and alcohol consumption

It was previously thought that most Sámi men were smokers (Eriksson *et al.*, 1970), but the reindeer herders' survey in 1986 found only 34% of the Sámi men smoked daily, compared with 27% of the Finnish men. The stereotypic view has been that heavy drinking is typical of the Sámi; however, the above-noted survey found the ethnic difference to be more moderate, with Sámi men drinking 22 grams of alcohol per day as compared with 12 grams among Finnish men. The percentage of heavy drinkers (20 grams or more daily) was 34% and 19%, respectively, while the number of occasions of getting drunk is 35 times a year, with no difference between the groups. The slightly greater amount of alcohol consumed per occasion among the Sámi seems to be largely ceremonial, as heavy drinking is no longer accepted (Poikolainen *et al.*, 1992). No investigation on alcohol-related health problems among the Finnish Sámi exists; however, Norwegian Sámi reportedly have lower percentages of such problems than Norwegians (Hetta *et al.*, 1978).

### Nutrition

The Sámi diet has been reviewed in many investigations. It is characterized by high consumption of reindeer meat and low consumption of milk, bread and butter, which has been replaced by margarine; however, variation exists between different groups of the Sámi (Jokelainen *et al.*, 1963). In the past, the whole reindeer was eaten (Haraldson, 1983). It is generally felt that the diet of the Finnish Sámi is becoming more similar to that of Finns, but there is no recent dietary survey based on the general population. In 1986, the Sámi reindeer herdsmen still showed dietary patterns different from those of their Finnish counterparts (Näyhä *et al.*, 1994; Laitinen *et al.*, 1996). The Sámi men ate reindeer meat four times a week and Finnish men twice a week, with respective amounts of 194 and 65 grams a day.

Other significant differences were the lower consumption of cow's milk and dark bread in the Sámi. Fatty reindeer milk was consumed in the past (Haraldson, 1983). The formerly low consumption of butter is no longer evident in the Sámi, but they still consume more margarine than Finns. The Sámi are said to be heavy coffee drinkers, and still consume eight cups a day; however, their consumption levels are similar to those of Finnish herders (seven cups a day). The dietary pattern of the Sámi reflects a slightly higher intake of protein (20% versus 17% of dietary energy), a slightly lower intake of fat (39% versus 42% of energy), and a higher polyunsaturated/saturated fatty acid (P/S) ratio (0.25 versus 0.18) than the Finns.

### Environmental pollution

In the 1950s the Sámi's home district was subjected to radionuclide fallout from Russian nuclear explosions in Novaja Zemlja (Rahola *et al.*, 1988). Since the fallout was deposited on lichen, the principal food of reindeer, radioactive substances could have entered the food chain. However, subsequent measurements showed no significant health risk which could be attributed to radiation. Similar findings were announced after the Chernobyl accident in 1986.

Up to the 1980s, no other significant sources of pollution were known, and Lapland was considered largely unpolluted. With the advent of *glasnost* the environmental pollution from industrial plants in northwestern Russia (the Kola Peninsula) became known, thus prompting several clarifying surveys. Moss studies pointed to total environmental damage around Russian smelters (e.g., the Petchenga-Nickel metallurgical complex 40 km from the Finnish border), but the polluted area was very limited and there remains uncertainty of its effects on Finnish Lapland. The main pollutants are heavy metals such as cadmium and mercury which, especially in the presence of environmental acidification, may be passed on to the Arctic food chain.

Investigations on reindeer herdsmen indicate increased exposure to some heavy metals in northeastern Lapland, where pollution is supposed to be most pronounced. The safe levels of cadmium in the blood (<45 nmol/l) were exceeded in 10% of the herders in the northeast (5% elsewhere), and most cases exceeding the WHO limit of 90 nmol/l critical for renal damage were Sámi (Luoma *et al.*, 1992*a,b*). Blood pressure was higher in men having high concentrations of blood cadmium, and cadmium was high in men suffering diagnosed arterial hypertension (Luoma *et al.*, 1995*a*). These findings were unaffected by smoking, which is known to increase blood cadmium. Despite the obvious geographical trend, a causal connection between pollution and blood cadmium remains unclear. Cadmium is accumulated in reindeer kidneys and liver, and due to advice given to local people, none of the herders ate reindeer organs. Blood mercury was elevated in 10% of the men, but none of the values reached the toxic levels (Luoma *et al.*, 1992*a,b*).

## Cardiovascular diseases and their risk factors

### Serum lipids

Early reports suggested that cardiovascular diseases among the Sámi were rare (Haraldson, 1962). Consequently, the Sámi were subjected to intensive research in Finnish Lapland in the late 1960s (Lewin and Eriksson, 1970). Relevant findings were that serum total cholesterol, high-density lipoprotein cholesterol and triglycerides were equal in the Sámi and in the Finns (Björkstén *et al.*, 1975), a finding repeated in Norway among local Sámi and local Finns (Thelle *et al.*, 1976; Thelle and Førde, 1979; Westlund, 1981). The reindeer herders' survey in 1986 (Näyhä *et al.*, 1994) found even higher serum total cholesterol in the Sámi than Finns (6.92 versus 6.50 mmol/l), similar to a recent Norwegian study (Ringstad *et al.*, 1991). The adverse lipid pattern in the Sámi could be partly ascribed to their fatty diet (Haraldson, 1983).

### Blood pressure

Sundberg *et al.* (1975) found Sámi blood pressures to be relatively low. Subsequent research with Sámi men showed systolic pressure 6 mm Hg lower than Finnish herders (Näyhä *et al.*, 1994). Similar findings of low blood pressure have been documented among Inuits (Young, 1986) and Arctic populations in Russia (Nikitin *et al.*, 1988). Previously, interpretations of lower blood pressure in the Sámi suggested a slow age-related increase, indicative of low levels of psychosocial stress (Sundberg *et al.*, 1975). However, the current age trend is similar in both ethnic groups.

The reasons for the relatively low systolic pressure in the Sámi are unknown, but long-term hard physical work may play a role (Marti, 1992). Drinking water taken from lakes and creeks low in salt content (Haraldsson, 1983) may lower blood pressure, but no research has been done on this. In any case, the low blood pressure in the Sámi is insufficient to compensate for the extra risk related to high serum cholesterol and smoking. The risk for coronary heart disease, based on smoking, blood pressure and serum total cholesterol was estimated to be a fifth higher in the Sámi than Finnish men (Näyhä and Hassi, 1993).

### Genetics: apolipoprotein E and A-IV polymorphism in the Sámi and the Finns

Apolipoprotein E (apoE) plays a central role in the metabolism of serum total cholesterol and triglycerides, and is a predictor for coronary artery disease (Utermann *et al.*, 1984) and Alzheimer's disease (Lehtimäki *et al.*, 1995). It is a structural constituent of chylomicrons, very low-density lipoproteins (VLDL) and high-density lipoproteins (HDL). ApoE is a ligand for the low-density lipoprotein (LDL) receptor and apoE receptor and mediates the uptake of its carrier lipoproteins by the liver (Brown and Goldstein, 1986). Three major apoE isoforms, apoE2, apoE3 and apoE4 can be distinguished by isoelectric focusing. The biosynthesis of apoE isoforms is controlled by three codominant alleles, $\varepsilon2$, $\varepsilon3$ and $\varepsilon4$. These alleles code for three apoE isoforms, E2, E3 and E4, which results in six apoE phenotypes: E2/2, E3/2, E4/2, E3/3, E4/3 and E4/4 (Utermann *et al.*, 1977). The apoE allele $\varepsilon4$ is associated with a high concentration of serum total and LDL cholesterol (Utermann, 1977). High frequencies of apoE allele $\varepsilon4$ have been reported in subjects with coronary artery disease (Cumming and Robertson, 1984, Utermann *et al.*, 1984).

We have compared the distributions of the apoE allele and phenotypes between the two ethnic groups, the Sámi and the Finns (Lehtinen *et al.*, 1996). ApoE allele frequencies in the Sámi and Finns were $\varepsilon2$: 0.051 versus 0.035; $\varepsilon3$: 0.645 versus 0.788; and $\varepsilon4$: 0.304 versus 0.177. The Sámi have a higher frequency of the $\varepsilon4$ allele than the Finns and most other populations. High $\varepsilon4$ allele frequencies have also been found in New Guineans, Bushmen, Nigerians, Sudanese and Blacks living in the USA, while low frequencies have been found in Icelanders and Norwegians (see Lehtinen *et al.*, 1996). The high occurrence of the $\varepsilon4$ allele in the Sámi may contribute to their high cholesterol, but this should increase rather than reduce their risk of coronary heart disease. The Sámi living in northern Norway do have higher serum cholesterol, but have a lower occurrence of coronary heart disease than Finns living in the same region (e.g., The Cardiovascular Study in Finnmark 1974–75, 1979).

Apolipoprotein A-IV (apoA-IV) is synthesized in the small intestine during fat absorption, and is a major component of newly synthesised triglyceride-rich lipoproteins derived from the intestine (Green *et al.*, 1980). It may be involved in the metabolism of chylomicrons and HDL, function as a cofactor for the enzyme lecithin cholesterol acyltransferase and promote efflux of cholesterol from cells (Steinmetz and Utermann, 1985). ApoA-IV is characterized by a genetically determined polymorphism, with isoforms identified by isoelectric focusing.

In Finland two common isoforms (A-IV-1 and A-IV-2) of apoA-IV are found. Expression of these most common codominant apoA-IV alleles results in three apoA-IV phenotypes: A-IV-1/1, A-IV-2/1 and A-IV-2/2 (Lukka *et al.*, 1988). Studies report populations in which the apoA-IV alleles influence plasma HDL cholesterol and triglyceride levels in such a way that subjects with the apoA-IV-2/1 phenotype have higher HDL cholesterol and lower triglyceride levels than those with the apoA-IV-1/1 phenotype (Menzel *et al.*,1988). However, the association between apoA-IV phenotype and HDL-cholesterol and triglyceride levels does not occur in all populations studied (Ehnholm *et al.*, 1994).

We have determined the frequencies of the two most common apoA-IV alleles in Sámi and Finnish reindeer herders and related apoA-IV phenotypes to serum lipid levels (Lehtinen *et al.*, 1998). ApoA-IV allele frequencies in the Sámi and the Finns were A-IV-1: 0.894 versus 0.944 and A-IV-2: 0.106 versus 0.056, respectively. The effect of apoA-IV-phenotype on serum HDL cholesterol levels differed significantly between the Sámi and the Finns. Sámi HDL cholesterol levels were significantly higher in apoA-IV-2/1 than in apoA-IV-1/1 phenotypes. The present study demonstrates that the Finns and the Sámi have different distributions of the most common apolipoprotein A-IV isoforms and phenotypes, with the Sámi having a higher frequency of the apoA-IV-2 allele than most European populations studied so far. Our results confirm the earlier reports that the apoA-IV-2 allele affects HDL cholesterol levels in some isolated populations, such as Icelanders and the Sámi. The effect of the apoA-IV phenotype on serum HDL cholesterol might be explained by a linkage disequilibrium between the apoA-IV gene locus and another locus or loci, which have direct physiological effects on plasma lipid levels. The gene for apoA-IV is located in chromosome 11 in band 11q23 closely linked to genes coding for apoA-I and C-III (Tenkanen and Ehnholm, 1992). There might be a locus within these genes, which interacts with codon 360 allele, and causes apoA-IV-2 allele to raise HDL cholesterol in the Sámi but not in the Finns.

### Occurrence of cardiovascular diseases

Haraldson (1962) reported that cardiovascular conditions were rare in Swedish Sámi compared with Swedes, and a 1970s Norwegian study found lower prevalence of self-reported myocardial infarction in Norwegian Sámi than in Norwegian Finns and Norsemen (Westlund, 1981). A subsequent Swedish mortality study based on reindeer owners found no significant difference in coronary heart disease between the Sámi and the general population of Sweden (Wiklund *et al.*, 1991). In Finland, the initial impression, in conjunction with research from the WHO Human Adaptability Program carried out in the 1960s, was that electrocardiographic changes indicative of coronary heart disease would be extremely rare in the Sámi (Lewin and Eriksson, 1970). However, a subsequent study found the prevalence of these abnormalities to be similar to that in Finns (Sundberg *et al.*, 1983). The reindeer herders' health survey cited above found prevalences of 8% and 12% for diagnosed heart disease for Sámi and Finns, respectively, and 2% and 4%, respectively, for diagnosed myocardial infarction, differences too small to reach statistical significance (Näyhä and Hassi, 1993). No other investigations comparing the occurrence of cardiovascular diseases between these two ethnic groups have been performed in Finland, but a recent report compared mortality from coronary heart disease between areas with different proportions of Sámi in their population.

The low mortality from coronary heart disease in the far north of Finland has been known for some time (Näyhä, 1989) and has been reported in detail, together with an analysis of recent trends (Näyhä, 1997). The Sámi district of Finland comprises the three northernmost government areas with a total population of approximately 10 000, of which 3500 are Sámi. The northernmost area (Utsjoki) is predominantly (80%) and the other areas partially Sámi (Inari 33%, Enontekiö 20%). During the period 1961–90, mortality from coronary heart disease in the predominantly Sámi Utsjoki was 41% and 47% lower in males and females, respectively, than in an adjoining, purely Finnish reference area. Similar but smaller differences were found for partially Sámi areas. The diagnoses entered on the death certificates were based on autopsy in only 14% of the cases; however, validity studies point to sensitivities and positive predictive values of 85% or more when using this diagnostic class (e.g., Stenbäck, 1986). Therefore, we consider these findings to be reliable. Moreover, mortality from all natural causes showed similar trends, which cannot be attributed to errors in death certification. The data broken down to sex-, age- and area-specific strata and subjected to multivariate analysis showed an association

RR

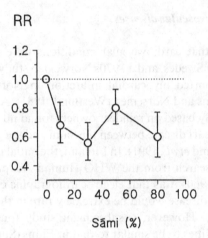

Sámi (%)

Figure 10.2. Age- and sex-adjusted risk ratios (RR) for mortality from coronary heart disease in 1961–90 and the percentage of Sámi in the population in 1970 (from Näyhä, 1997).

between coronary heart disease and the percentage of Sámi in the population (Figure 10.2). However, this correlation was not entirely consistent, mainly because strata having 40–60% Sámi (this was based on Sámi men over 55 years old living in Inari) showed relatively "high" mortality. This was interpreted as a result of more advanced urbanization in the Inari area. It was surprising that during the period 1961–90 mortality from coronary heart disease in Utsjoki, the core Sámi area, was declining both in absolute terms and relative to the Finnish reference area (Figure 10.3).

Although coronary heart disease is claimed to be rare among the Sámi, investigations performed so far are unable to document this. One difficulty is in defining the Sámi ethnicity. The population of Finland contains Sámi admixture to varying extents (Eriksson, 1973). Consequently, it may be impossible to create an unbiased study design to compare the two ethnic groups. Additionally, lifestyle factors may play a role, especially since the local diet is partly shared by both ethnic groups.

### Proposed protective factors for coronary heart disease

Since the classic risk factors for coronary heart disease among the Sámi point to a high rather than a low risk, other factors, possibly both genetic and environmental, may be responsible for the low occurrence of this condition among them. Several investigations provide evidence that anti-

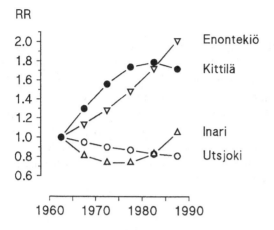

Figure 10.3. Changes in mortality from coronary heart disease in 1961–90. Regression-based risk ratios by area (1.00 = mortality in 1961–65) (from Näyhä, 1997).

oxidants may play a role in the prevention of atherosclerosis and athero-sclerotic vascular disease. Variation in consumption of antioxidants such as alpha-tocopherol (vitamin E; Gey and Puska, 1989), retinol (vitamin A; Gey and Puska, 1989), albumin (Phillips *et al.*, 1989) and selenium (Salonen *et al.*, 1982) could explain the cross-cultural differences in the incidence of coronary heart disease better than the classic risk factors (Gey *et al.*, 1994). Our earlier studies have revealed high serum alpha-tocopherol, albumin and selenium levels in men living in the Sámi area of Finland, which also records a low mortality from coronary heart disease (Luoma *et al.*, 1995*b*, 1996; Näyhä, 1997). The men living in communities with high coronary heart disease mortality showed low serum alpha-tocopherol and albumin levels, and *vice versa* (Luoma *et al.*, 1995*a*, 1996). Serum alpha-tocopherol increased with the consumption of reindeer meat, and serum selenium increased with fish consumption (Luoma *et al.*, 1995*b*). The good antioxidant status may be a factor in the anomalous, low mortality from coronary heart disease in people with high serum choles-terol living in the Sámi area.

## Cancer

Information on the occurrence of cancer among the Sámi is restricted to Sweden, where the radionuclide fallout in the 1950s was suspected to cause an increase in cancer incidence (Wiklund et al., 1989). Although the cesium-137 content in reindeer meat had increased early in the 1960s, and elevated contents had been measured in the Sámi, the incidence of cancer, including that of breast, thyroid and bone marrow, which are regarded to be especially susceptible to radiation, was low. The single exception was cancer of the stomach, with the incidence potentially attributed to a high intake of salt and smoked food.

## Conclusions

The Sámi of Finland are undergoing an epidemiological transition, with the receding of infectious diseases and emergence of degenerative diseases. The information on the occurrence of "modern" diseases among the Sámi is deficient. The cancer pattern of the Finnish Sámi remains unknown, but the Swedish experience points to a low occurrence of most cancers, with the exception of cancer of the stomach, which is held to be typical of poor living conditions. Current experience points to a low occurrence of coronary heart disease, irrespective of high levels of risk factors such as smoking and serum cholesterol. The high serum cholesterol among the Sámi might be partly explained by the high frequency of apoE allele $\varepsilon 4$, which is associated with high concentration of serum total and LDL cholesterol. Low blood pressure and dietary antioxidants are hypothesized as protective factors against atherosclerosis and coronary heart disease. Epidemiological studies of coronary heart disease among the Sámi would be highly useful, especially in Finland, which otherwise suffers from a high occurrence of this disease.

## References

Björkstén, F., Aromaa, A., Eriksson, A.W., Maatela, J., Kirjarinta, M., Fellman, J. and Tamminen, M. (1975). Serum cholesterol and triglyceride concentrations of Finns and Finnish Lapps. II. Interpopulation comparison and occurrence of hyperlipidemia. *Acta Medica Scandinavica* **198**, 23–33.
Beach, H. (1988). *The Saami of Lapland.* The Minority Rights Group Report No. 65. London: Minority Rights Group.

Brown, M.S. and Goldstein, J.L. (1986). A receptor-mediated pathway for cholesterol homeostastis. *Science* 232, 34–47.

The Cardiovascular Study in Finnmark 1974–1975. (1979). *Nordic Council for Arctic Medical Research Report* 25, 1–195.

Cumming, A.M. and Robertson, F.W. (1984). Polymorphism at the apolipoprotein E locus in relation to risk of coronary disease. *Clinical Genetics* 25, 310–313.

Ehnholm, C., Tenkanen, H., de Knijff, P., Havekes, L., Rosseneu, M., Menzel, H.J.X. and Tiret, L. (1994). Genetic polymorphism of apolipoprotein A-IV in five different regions of Europe. Relations to plasma lipoproteins and to history of myocardial infarction: the EARS study. European Atherosclerosis Research Study. *Atherosclerosis* 107, 229–238.

Eriksson, A.W. (1973). Genetic polymorphism in Finno-Ugrian populations. *Israel Journal of Medical Sciences* 9, 1156–1170.

Eriksson, A.W., Fellman, J., Forsius, H. and Lehman, W. (1970). Phenylthiocarbamide tasting ability among Lapps and Finns. *Human Heredity* 20, 623–630.

Gey, K.F. and Puska, P. (1989). Plasma vitamins E and A inversely correlated to mortality from ischaemic heart disease in cross-cultural epidemiology. *Annals of the New York Academy of Sciences* 57, 268–282.

Gey, K.F., Stähelin, H.B. and Ballmer, P.E. (1994). Essential antioxidants in cardiovascular diseases: lessons for Europe. *Therapeutische Umschau* 51, 475–482.

Green, P.H., Glickman, R.M., Riley, J.W. and Quinet, E. (1980). Human apolipoprotein A-IV. Intestinal origin and distribution in plasma. *Journal of Clinical Investigations* 65, 911–919.

Haraldson, S. (1962). Levnads- och dödlighetsförhållanden i de nordligaste svenska lappbyarna. *Svenska Läkartidningen* 59, 2829–2844.

Haraldson, S.R.S. (1983). Health and disease among the Lapps. *Polar Record* 21, 345–357.

Hetta, O.M., Keskitalo, A.I., Somby, P., Vandbakk, Ø., Nordsletta, A. and Solbakk, R. (1978). Views on alcoholism in the Sami areas of Finnmark. *Nordic Council for Arctic Medical Research Report* 21, 42–48.

Jokelainen, A., Pekkarinen, M., Roine, P. and Miettinen, J.K. (1963). The diet of Finnish Lapps. *Zeitschrift für Ernährunswissenschaft* 3, 110–117.

Laitinen, J., Näyhä, S., Sikkilä, K. and Hassi, J. (1996). Diet and cardiovascular risk factors among Lapp and Finnish reindeer herders. *Nutrition Research* 16, 1083–1093.

Lehtimäki, T., Pirttilä, T., Mehta, P.D., Wisniewski, H.K., Frey, H. and Nikkari, T. (1995). Apolipoprotein E (apoE) polymorphism and its influence on ApoE concentrations in the cerebrospinal fluid in Finnish patients with Alzheimer's disease. *Human Genetics* 95, 39–42.

Lehtinen, S., Lehtimäki, T., Luoma, P., Näyhä, S., Ehnholm, C. and Nikkari, T. (1996). Differences in allele distribution of apolipoprotein E between Lapps and Finns. *European Journal of Laboratory Medicine* 4, 39–44.

Lehtinen, S., Luoma, P., Näyhä, S., Hassi, J., Ehnholm, C., Nikkari, T., Peltonen, N., Jokela, H., Koivula, T. and Lehtimäki, T. (1998). Apolipoprotein A-IV polymorphism in Saami and Finns: frequency and effect on serum lipid levels. *Annals of Medicine* **30**, 218–223.

Lewin, T. and Eriksson, A.W. (1970). The Scandinavian International Biological Program, Section for Human Adaptability, IBP/HA. Scandinavian IBP/HA investigations in 1967–1969. *Arctic Anthropology* **7**, 63–68.

Lewin, T. and Hedegård, B. (1971). Human biological studies among Skolts and other Lapps. A survey of earlier literature. *Transactions of the Finnish Odontological Association* **67**(Suppl. 1), 64–70.

Lewin, T., Hedegård, B. and Kirveskari, P. (1971). Odontological conditions among the Lapps in Northern Fenno-Scandia. *Transactions of the Finnish Odontological Association* **67**(Suppl. 1), 99–103.

Lewin, T., Rundgren, Å., Louekari, L. and Forsius, H. (1971). Child mortality among the Skolt Lapps. *Transactions of the Finnish Odontological Association* **67**(Suppl. 1), 39–54.

Lukka, M., Metso, J. and Ehnholm, C. (1988). Apolipoprotein A-IV polymorphism in the Finnish population: gene frequencies and description of a rare allele. *Human Heredity* **38**, 359–362.

Luoma, P.V., Näyhä, S., Korpela, H., Pyy, L. and Hassi, J. (1992a). Blood mercury and serum selenium concentrations in reindeer herders in the arctic area of northern Finland. *Archives of Toxicology* Suppl. 15, 172–175.

Luoma, P.V., Näyhä, S., Pyy, L., Korpela, H. and Hassi, J. (1992b). High blood cadmium and mercury concentrations in reindeer herders in northern Finland west of Kola Peninsula. In *Symposium on the State of the Environment and Environmental Monitoring in Northern Fennoscandia and the Kola Peninsula*, ed. E. Tikkanen, M. Varmola and T. Katermaa, pp. 349–351. Rovaniemi, Finland: Arctic Centre Publications 4.

Luoma, P., Näyhä, S., Pyy, L. and Hassi, J. (1994). Blood cadmium and blood pressure in reindeer herders living in northern Finland. *Arctic Medical Research* **53**, 376–377.

Luoma, P.V., Näyhä, S., Pyy, L. and Hassi, J. (1995a). Association of blood cadmium to the area of residence and hypertensive disease in arctic Finland. *Science of the Total Environmen* **160/161**, 571–575.

Luoma, P.V., Näyhä, S., Sikkilä, K. and Hassi, J. (1995b). High serum alpha-tocopherol, albumin, selenium and cholesterol, and low mortality from coronary heart disease in northern Finland. *Journal of Internal Medicine* **237**, 49–54.

Luoma, P.V., Näyhä, S., Sikkilä, K. and Hassi, J. (1996). Antioxidants, diet and mortality from ischaemic heart disease in rural communities in northern Finland. In *Natural Antioxidants and Food Quality in Atherosclerosis and Cancer Prevention*, ed. J.T. Kumpulainen and J.T. Salonen, pp. 123–129. London: The Royal Society of Chemistry.

Marti, B. (1992). Physische Aktivität und Blutdruck. Eine epidemiologische Kurzreview des primärpreventiven Effekts von körperlich-sportliche Betätigung. *Schweizerische Rundschau für Medizinische Praxis* **81**, 473–479.

Menzel, H.J., Sigurdsson, G., Boerwinkle, E., Schrangl-Will, S., Dieplinger, H.X. and Utermann, G. (1988). Frequency and effect of human apolipoprotein

A-IV polymorphism on lipid and lipoprotein levels in an Icelandic population. *Human Genetics* **84**, 344–346.

Middaugh, J. (1990). Cardiovascular deaths among Alaskan Natives, 1980–86. *American Journal of Public Health* **80**, 282–285.

Näyhä, S. (1989). Geographical variations in cardiovascular mortality in Finland, 1961–1985. *Scandinavian Journal of Social Medicine*, Suppl. 40.

Näyhä, S. (1997). Low mortality from ischaemic heart disease in the Sámi district of Finland. *Social Science & Medicine* **44**, 123–131.

Näyhä, S. and Hassi, J., ed. (1993). *Life style, work and health of Finnish reindeer herders* (ML 127). Finland: Social Insurance Institution.

Näyhä, S., Sikkilä, K. and Hassi, J. (1994). Cardiovascular risk factor patterns and their association with diet in Saami and Finnish reindeer herders. *Arctic Medical Research* **53**, 301–304.

Nikitin, Yu., Asthakhova, T., Khasnulin, V., Serova, N., Gafarov, V., Ukolova, L., Filimonova, T., Sheludko, L. and Bulgakov, Yu. (1988). Arterial hypertension and risk factors of CHD in native and newcoming populations of some regions of Siberia and the far north. *Arctic Medical Research* **47**(Suppl. 1), 469–471.

Palva, I.P. and Finell, B. (1976). Tuberculosis in Finnish Lapland. In *Circumpolar Health. Proceedings of the 3rd International Symposium, Yellowknife, NWT*, ed. R.J. Shephard and S. Itoh, pp. 331–334. Toronto: University of Toronto Press.

Phillips, A., Shaper, A.G. and Whincup, P.W. (1989). Association between serum albumin and mortality from cardiovascular diseases, cancer, and other causes. *Lancet*, **334**, 1434–1436.

Poikolainen, K., Näyhä, S. and Hassi, J. (1992). Alcohol consumption among male reindeer herders of Lappish and Finnish origin. *Social Science & Medicine* **35**, 735–738.

Rahola, T., Jaakkola, T., Miettinen, J.K., Tillander. M. and Suomela, M. (1988). Radiation dose to Finnish Lapps: comparison of effects of fallout from atmoshperic nuclear weapons tests and from the Chernobyl accident. *Arctic Medical Research* **47**(Suppl. 1), 186–190.

Ringstad, J., Aaseth, J., Johnsen, K., Utsi, E. and Thomassen, Y. (1991). High serum selenium concentrations in reindeer breeding Lappish men. *Arctic Medical Research* **50**, 103–106.

Salonen, J.T., Alfthan, G., Huttunen, J.K., Pikkarainen, J. and Puska, P. (1982). Association between cardiovascular death and myocardial infarction and serum selenium in a matched-pair longitudinal study. *Lancet* **II**, 175–179.

Sing, C.F. and Davignon, J. (1985). Role of the apolipopotein E polymorphism in determining normal plasma lipid and lipoprotein variation. *American Journal of Human Genetics* **37**, 268–283.

Stenbäck, F. (1986). Accuracy of antemortem diagnosis in the north. An autopsy study. *Arctic Medical Research* **41**, 9–15.

Steinmetz, A. and Utermann, G. (1985). Activation of lecithin: cholesterol acyltransferase by human apolipoprotein A-IV. *Journal of Biological Chemistry* **260**, 2258–2264.

Sundberg, S., Luukka, P., Lange Andersen, K., Eriksson, A.W. and Siltanen, P. (1975). Blood pressure in adult Lapps and Skolts. *Annals of Clinical Research* 7, 17–22.

Sundberg, S., Siltanen, P., Lange Andersen, K., Kirjarinta, M., Lietzén, R. and Sahi, T. (1983). Resting- and postexercise ECG findings in an adult Lapp population. *Annals of Clinical Research* 15, 50–54.

Tenkanen, H. and Ehnholm, C. (1992). Molecular basis for apo A-IV polymorphisms. *Annals of Medicine* 24, 369–374.

Thelle, D.S. and Førde, O.H. (1979). The Cardiovascular Study in Finnmark County: coronary risk factors and the occurrence of myocardial infarction in first degree releatives and in subjects of different ethnic origin. *American Journal of Epidemiology* 110, 708–715.

Thelle, D.S., Førde, O.H., Try, K. and Lehmann, E.H. (1976). The Tromsˇ Heart Study. Methods and main results of the cross-sectional study. *Acta Medica Scandinavica* 200, 107–118.

Utermann, G. (1987). Apolipoprotein E polymorphism in health and disease. *American Heart Journal* 113, 433–440.

Utermann, G., Hees, M. and Steinmetz, A. (1977). Polymorphism of apolipoprotein E and occurrence of dysbetalipoproteinemia in man. *Nature* 269, 604–607.

Utermann, G., Hardewig, A. and Zimmer, F. (1984). Apolipoprotein E phenotypes in patients with myocardial infarction. *Human Genetics* 65, 237–241.

Westlund, K. (1981). Circumpolar non-infectious epidemiology. *Nordic Council for Arctic Medical Research Report* 31, 215–220.

Wiklund, K., Holm, L.E. and Eklund, G. (1989). Låga cancerrisker bland svenska renskötande samer. *Läkartidningen* 86, 2841–2844.

Wiklund, K., Holm, L.E. and Eklund, G. (1991). Mortality among Swedish reindeer breeding Lapps in 1961–85. *Arctic Medical Research* 50, 3–7.

Young, T.K. (1986). Epidemiology and control of chronic diseases in circumpolar Eskimo/Inuit populations. *Arctic Medical Research* 42, 25–47.

Young, T.K. (1988). Are subarctic Indians undergoing the epidemiologic transition? *Social Science & Medicine* 26, 659–671.

# 11 *Yomut family organization and demography*

WILLIAM IRONS

## Introduction

The family organization of the Yomut Turkmen subordinates the interests of women to those of men. This is reflected in their household organization which emphasizes keeping closely related males together in the same household and requires women to move from their natal households to their husbands' households after marriage. Among the Yomut, men stay in an extended family headed by their fathers for a relatively long period after marriage. Women thus spend most of their reproductive years as daughters-in-law in the household of their parents-in-law. The norms of social interaction within a household place them in a position of subordination to their husbands and parents-in-law and entail a degree of hardship. The patrilateral bias of Yomut family organization appears to be economically advantageous even though it imposes a cost on women married into the household. This economic advantage is tied closely to the Yomut sexual division of labor which assigns agricultural and pastoral labor to men and assigns women work near the home base that is compatible with child care.

The disadvantageous situation of women in Yomut society appears to be reflected in higher death rates for Yomut women during their reproductive years. These are the years during which their status within their households is at its lowest. A woman enjoys a higher status (though lower than that of men) while still resident in her father's household, and again later in life when her husband is an independent household head and she is superordinate to her daughters-in-law within her husband's household. During these years, for the most part, mortality rates for women are not significantly different from those of men. In the first 2 years of life, females have a lower mortality rate than males.

These differences in death rates need to be interpreted in light of the fact that in most human populations males have higher death rates at all ages than do females. This condition generally characterizes species with a history of polygynous breeding, and most human populations fit this

251

pattern. In the case of human populations, however, death rates are influenced by both universal biological characteristics of our species and by locally variable culture. In the case of the Yomut, culture appears to override the biologically based greater vulnerability of males and creates a social environment in which women have higher death rates than men for a significant portion of their life histories, specifically during their reproductive years.

I suggest that in situations of social equality, men will have higher death rates at all ages than women because of their biologically greater vulnerability. I further suggest that the extent to which a population deviates from a higher male death rate can be taken as a probable indicator that women are socially disadvantaged.

### The Yomut

The Yomut are one of several large Turkmen descent groups, which along with the Teke, Goklan, Ersari, Salor and Sariq, constitute the majority of the Turkmen ethnic group. These descent groups occupy a contiguous area in what is now the Islamic Republic of Turkmenistan (the former Turkmen Soviet Socialist Republic) and adjacent areas of Iran and Afghanistan. The Yomut in particular occupy two regions within this larger Turkmen area. One region consists of an area in northern Iran know as the Gorgan Plain and adjacent areas of the Republic of Turkmenistan. In Iran, this region is also referred to as the Turkmen Plain. The second region is to the north in the vicinity of the city of Khiva.

My informants in northern Iran defined Turkmen as people who met three criteria. First, they had to speak Turkmen as their mother tongue. Second, they had to be descended either from Turkmen of earlier generations or slaves of the Turkmen of earlier generations. Third, they had to be Sunni Moslems of the Hanafi rite. The core of groups such as the Yomut, Teke, and Goklan are all conceived of as descendants of a single ancestor in the distant past, a man named Oghuz Khan. They have elaborate genealogies tracing the ancestry of groups like the Yomut from this single ancestor. The Turkmen ethnic group also includes certain sacred descent groups known as the Ewlad. These groups are thought to have resided among the Turkmen for a very long time but ultimately to be descendants of the first four Khalifs, and thus to be of Arab origin. Other smaller groups such as the Nokhorli and the Yemreli were included among the Turkmen as groups entirely of slave origin. By this definition, numerous groups in the Middle East and numerous historic groups that referred to themselves as Turkmen would be excluded. To avoid confusion, one can refer to the Turkmen as defined by my informants as the Central Asian Turkmen. The language of the Central Asian Turkmen is related to Turkish, but the two languages are not mutually comprehensible when speakers of the two languages first encounter one another. With extended exposure, however, speakers of these two languages begin to understand one another (cf. Hockett, 1958).

The Yomut are a largely endogamous population. They prefer cousin marriage and local community endogamy. Roughly 90% of Yomut men marry Yomut women. The Yomut never allow Yomut women to marry outside the Yomut, but a small percentage of Yomut men take wives from groups such the Goklan Turkmen, the Ewlad Turkmen, and various non-Turkmen groups that reside near Yomut territory (Irons, 2000). The Yomut consider these other groups from which they occasionally take wives to be of lower status. This pattern of marriage in which women can marry men from their own group or higher-status groups, but can not marry men from lower status groups is known as hypergyny and is common in many parts of the world where there are stratified descent and ethnic groups.

### Economy

The Yomut of the Gorgan Plain consciously divide themselves into two groups: the *chomur* and the *charwa*. The *chomur* are predominantly agri-cultural and the *charwa* are predominantly pastoral. However, this is a difference in emphasis and both groups practice both agriculture and pastoralism. Both groups are rural producers who practice a combination of subsistence production and production for market exchange. The *charwa* predominantly raise sheep and goats. The agricultural Turkmen traditionally lived primarily by rainfall cultivation of wheat and barley along with some herding of sheep and goats. More recently many of the *chomur* have become cultivators of cotton for markets. The more agricul-tural Yomut, the *chomur,* appear to be derived historically from more pastoral Yomut, the *charwa* (Barthold, 1962). Until the forced sedentariz-ation of Reza Shah, the *chomur* lived in yurts and made periodic short migrations. Like the *charwa* they were also active raiders of neighboring communities. At the time of my research, the family organization and the sexual division of labor among the *chomur* were not different from that of the *charwa* even though they had assumed a more sedentary residence pattern. The Yomut as a whole are divided into a hierarchy of smaller and smaller descent groups based in genealogy. Most of these smaller descent groups collectively owned a territory that included areas suitable to both the *chomur* and *charwa* economies, and commonly there was considerable movement of population between these two groups.

Since the sedentarization program of the 1930s, this situation has changed. During the period of government enforced sedentarization (roughly 1930 to 1941), the *chomur* live in mud houses, while the *charwa* spend part of the year living in mud houses and part of the year migrating in yurts. After the Russian occupation of northern Iran in 1941, the forced sedentarization program of Reza Shah ended and the *charwa* returned to full-time residence in yurts. Most of the *chomur*, however, continued to live in mud houses after the Russian occupation. Later in the 1970s, some *charwa* began voluntarily to combine part-time residence in mud houses with residence in yurts.

However, whatever their residence pattern, the *charwa* continued to make their living primarily by raising sheep and goats, with rainfall cultivation of wheat and barley as a secondary activity. Agriculture is limited to areas such as valley bottoms or natural depressions where excess of rainwater collects. Other parts of the *charwa* region are suitable only as pasture. The *charwa* also derive a portion of their income from raising horses and from carpet weaving using the wool of their sheep.

Since the 1950s, the *chomur* in the southern part of *chomur* territory where irrigation is possible have taken up commercial cotton cultivation and have become very prosperous by rural Iranian standards. In the drier northern portion of *chomur* territory, the Yomut derive their living primarily from rainfall cultivation of wheat. Trade and the full-time teaching of religion are also secondary occupations in both *chomur* and *charwa* communities.

Since the 1950s, some Yomut traders have taken up residence in the cities that have grown up inside traditional Yomut territory. These include the cities of Bandar Turkoman (formerly Bandar Shah), Pahlavi Dejh, Gonbad-e Kavus, Gomishan, and Maraveh Tappeh. By and large these newly urbanized Yomut are prosperous. They have not, however, changed their ideas about family organization, marriage and kinship.

At the time of my research, the Yomut on the whole were a prosperous group by rural Iranian standards. One result of this prosperity was that there was almost no migration of Yomut out of the Gorgan Plain. The southern portion of Yomut country, which is suitable for irrigation, became especially prosperous after the introduction of irrigated cotton production in the 1950s, and numerous non-Yomut migrated into the southern portion of Yomut country. Large grants of land in this southern region were given to various high-status Iranians during the reigns of Reza Shah and his son, Mohammed Reza Shah, and these were the sites of modern farms employing the labor of impoverished migrants from southeastern Iran, from Sistan, Zabol, and Baluchistan. At the time of my research, the modern farms and traditional *chomur* Yomut communities were interspersed in the southernmost part of traditional Yomut country. The northern portion of Yomut country including the drier portions of *chomur* country and the *charwa* region were still exclusively inhabited by Yomut plus scattered communities of Ewlad. The city of Gonbad-e Kavus was near the region where I did most of my fieldwork and there was a sizable Yomut population in this city. These urban Yomut were all fairly recent immigrants from the surrounding rural Yomut communities. Some, who were educated, had government jobs, and some were merchants. Rural Yomut when they went to Gonbad-e Kavus for trade commonly traded at the store of a member of their own descent group, and were commonly put up as overnight guests by the lineage-mate with whom they traded. Many of the urban Yomut were wealthy and maintained a house both in the city and in the rural community of their origin. Very few of them fit the stereotype of impoverished rural people entering the urban population at the bottom of the economic hierarchy.

### Political organization

The Yomut are traditionally an acephalous society (Irons, 1971, 1979, 1994). For purposes of defending territory and resisting government control, they were able to use their segmentary lineage system to raise large temporary military units (Irons, 1974). Such units would for the time of their existence elect a leader. Also, near the territory inhabited by sedentary non-Turkmen, there were men who organized the collection of tribute from these sedentary non-Turkmen. Aside from these positions of limited formal leadership, the Yomut had no chiefs. All household heads were considered equal and collective decisions by residence groups were made by discussion and consensus among the household heads of the group. Rights of person and property were protected only by threat of retaliation, not by any centralized authority. Descent groups were patrilineal and residence groups tended to correspond to descent groups. Homicides or theft of property were redressed by the patrilineal kin of the victims seeking revenge in the case of homicide or repayment in case of theft. The size of the groups mustered for this purpose depended on the importance of the issue being redressed. When disputes did arise, a rule of

256     W. Irons

complementary opposition applied and individuals were expected to side with those to whom they were most closely related in the male line. This meant that residing with close patrilineal kin was important for protecting one's rights of person and property.

After the conquest of the Turkmen by the Iranian army under the leadership of Reza Shah in 1925, efforts to establish effective government control of the Yomut were put in place. However, these efforts ceased in 1941, when the Soviet army occupied northern Iran and allowed the Yomut to regulate their own affairs so long as they did not interfere with the transportation of military supplies through their territory to the Soviet Union. Thus the transition from an acephalous society regulated by a balance of threats of violence between patrilineal descent groups to a society with government enforced law and order was reversed in 1941. Following the end of World War II, the Iranian government gradually reasserted control of the internal affairs of the Yomut. Once the government asserted its right to enforce law and order among the Yomut, the descent groups continued to play a role in internal affairs, to a degree, both by using violence without government sanction and also by raising money for bribery along lines of patrilineal descent. Thus at the time of my research patrilineal kin were important at every level of Yomut society. Households that were closely related in the male line would commonly gather their men (or their money) together to confront similar groups as a means of resolving disputes. This situation tended to reinforce the emphasis on patrilineal kinship among the Yomut.


#### Education

Before the reform and modernization efforts of Reza Shah in the 1930s, the only form of formal education found among the Yomut was religious education. Among the Yomut there were men who were considered qualified to teach religion, and as part of the training they taught their students to read Arabic, the language of the Quran and of numerous books relevant to Islam. These men were known as akhonds, and were addressed with the word akhond before their names as an honorific. These men often taught religion out of their yurts. Occasionally especially successful akhonds would have a mosque and school built including simple dormitories where their students could live while studying. The more influential of such akhonds often were also leaders (*pirs*) of a Sufi order and combined the teaching of religion with the teaching of a particular Sufi discipline. Akhonds and some Sufis distinguished themselves by wearing a turban rather that the traditionally black karakul hat of the Turkmen. Men who studied at the feet of an akhond without learning enough to become akhonds themselves were known as mullahs. Mullahs varied in the extend of their literacy. Some had learned very little, others were very literate. Those who had considerable learning also carried the honorific mullah before their names. Mullahs assumed responsibility for teaching children the basics of their religion.

In the *charwa* group with which I first lived in 1966, about 2% of the population were literate as a result of this traditional form of education. Those who had learned to read and write Arabic could quickly transfer this skill to a reading of Turkmen literature written with the Perso-Arabic alphabet. They also as a rule learned to speak, read, and write Persian (also known as Farsi), the national language of Iran. These literate individuals frequently served as government-appointed headmen for local communities because of their literacy and knowledge of Persian.

As part of the reform program of Reza Shah schools for Turkmen were established in the western part of Yomut territory in the newly established city of Bandar Shah (now Bandar Turkoman), a city which also served as a railhead for the Iranian railway built as part of Reza Shah's development program. During the "White Revolution" developed by Reza Shah's son, Mohammed Reza Shah, in the 1960s and later, there was an extensive effort to provide elementary education throughout rural Iran, and village schools were established and staffed by high-school graduates doing their obligatory military service. When my research began, most of the *chomur* communities had government elementary schools staffed with these high school graduates. As a result a large portion of the children and adolescents among the *chomur* were literate, but the majority of the adult population were still not literate unless they had had a religious education. In the Yomut region, the effort to establish government schools in rural areas began in the more prosperous *chumur* communities and only gradually spread to remoter, less prosperous communities. In the *charwa* community where I began my first participant observation in 1966, the first government school was set up in a yurt in 1973.

The overwhelming majority of those educated in the new government schools were male. Of the roughly 100 Yomut communities that my wife and I visited in 1973 to establish a sample frame for our survey, only four had girls' schools. The four girls' schools we did see were in large villages and were staffed by female high-school graduates drawn from the urban Turkmen population. In other villages, occasionally girls were allowed to attend the first few years of school in a mixed class, but this was the exception rather than the rule. The Yomut do not see the mixing of boys and girls in the same school as suitable for other than very young children. They also see education for women as something of lesser value than for men. Further, there was some opposition to education for females among some of the more influential akhonds.

Among urban Turkmen, education often went further, and many urban Turkmen received a high-school education. A very small number went on to university. Also, prosperous rural families would sometime send their sons to live with relatives in the city of Gonbad-e Kavus so they could acquire an education beyond elementary school. Education in the government schools was in Persian, and thus gradually the rural Turkmen were being made literate and were learning the national language. Again, however, the situation varied from one region to another. In the community where I did my first work, there were only two men, both mullahs, with whom I could carry on a conversation in Persian. A handful of other men knew a few words of Persian. The main language of the community was Turkmen with the consequence that I was forced to learn that language.

For a fuller description of the Yomut of the Gorgan Plain see Irons (1975).

### The data

The data used for this chapter were gathered during three field trips to the Yomut region. The first period of field research was from December 1965 to November 1967 and included 16 months of actual residence with the Yomut, most of it in a single *charwa* community (Irons, 1975). The second was for 2 months in the summer of 1970 spent in the same *charwa* community. The third period of field research was from August of 1973 to August of 1974. During this period of research my wife, Marjorie Rogasner, and I lived in a *chomur* community and conducted a survey of a large sample of Yomut households in both *chomur* and *charwa* regions. The sample frame for the survey was constructed by visiting all of the communities of two subgroups of the Yomut, the Qojuq and the Ighdar. With the assistance of a sampling expert, Alan Ross of Johns Hopkins University, we then drew a sample of 21 communities and from each village a sample of roughly 25 households. In a few cases, somewhat larger samples were taken and the final sample consisted of 566 households. Also several Yomut households in the city of Gonbad-e Kavus were included in the sample. These were members of the Qojuq descent group. Most of the Qojuq residents in Gonbad-e Kavus were prosperous families who maintained houses both in Gonbad-e Kavus and in the Qojuq community of their origin, which in most cases were within 10–20 miles of the city.

The sample was a random stratified sample designed to maximize the opportunities to detect variation in demographic parameters within the Yomut population rather than to maximize the accuracy of the estimates of the overall population parameters.

The survey gathered data on household histories, wealth, and demographic history. This data has made possible estimates of Yomut birth and death rates and estimates of the effects of various social variables, especially wealth, on birth and death rates. Such data has been presented in a number of earlier publications (Irons, 1979, 1980, 1986, 1994, 2000). The demographic data presented in this paper derives from this survey and the ethnographic data are the product of the three field sessions mentioned above.

### Yomut family organizaiton

The Yomut Turkmen have a family organization that emphasizes the father–son relationship above all other family relationships and secondarily emphasizes the brother–brother relationship. Yomut households are built around a core of father–son and brother–brother relationships. Close female relatives, daughters and sisters, are expected to subordinate their interests first to their fathers and brothers and eventually to those of their husbands. Marriage for women is universal and the average age at marriage is 15 years. However, permanent residence with husbands is ideally delayed until 3 years after marriage, and typically is delayed for from 1 to 3 years.

A woman's status in her natal household is higher than that of a married-in daughter-in-law but lower than that of her brothers. Thus on assuming permanent residence with a husband, a Yomut woman assumes a very subordinate position in the household of her parents-in-law. Sons typically remain members of their father's households for a number of years after their wives take up residence with them, and usually do not establish independent households until they have children approaching the age of marriage.

For wealthy families the age of marriage for men is also usually around 15 years, but for poorer families, marriages for sons are often not arranged until they are somewhat older. Very poor men occasionally marry at an age close to 40 years, and, in the past when the general economic condition of the Yomut was poorer, some men never married.

### The developmental cycle of Yomut households

The developmental cycle of domestic groups begins when a son separates from his father's household and is given a patrimony in land and livestock from which to derive a living. After separation, a son usually continues to live near his father and interaction between the father's and the son's households is extensive. If they are nomadic the two households tend to migrate and camp together. The son is usually married and has children approaching maturity when he separates from his father, and over the years his sons marry and bring their wives into the household while the daughters marry and leave to join their husbands. Then the cycle repeats itself when the household head's sons eventually separate off, in turn, to form independent households. The youngest son does not separate off, but instead remains with his parents, cares for them in their old age, and eventually inherits the residual livestock and land holdings of the parents. The developmental cycle is so dependent on the presence of sons, that men without sons potentially face a difficulty. This is usually resolved by the adoption of a son from a close agnatic relative. Usually a brother's son is adopted. The adopted son then occupies the position of a youngest son for all economic purposes even though his place in the patrilineal genealogy remains that of his birth.

The developmental cycle can be seen as a process of accumulating wealth and then, in effect, converting the wealth into people. In the early stages of a household's life cycle, the household accumulates wealth and then uses this wealth to acquire brides for the family's sons, and eventually to allow sons to separate off, establishing independent households of their own. Both paying bridewealth and giving patrimonies to sons requires accumulating wealth beyond that which is necessary for maintaining a livelihood. When sons are married, the head of the household, usually the groom's father, has to pay a large bridewealth to the father and mother of the bride. This bridewealth is fixed by convention at ten camels for the father and one for the mother. However, the actual form of the bridewealth can, and usually is, commuted to a form other than camels. The formula for commuting the bridewealth allows the substitution of ten sheep or goats for each camel, and usually among the pastoral Yomut, at the time of my research, this was the form of payment. Among the agricultural Yomut, at the time of my research, the payment was usually commuted to the current cash value of the number of livestock required. Whatever form the bridewealth takes, it is fixed at the value equivalent to eleven camels, and it usually cannot be negotiated to a lower amount for poorer families. Occasionally when marriages are arranged between first cousins, the bridewealth is reduced as an expression of kin altruism. (In this case, the heads of the households arranging the marriage for their children are usually bothers.) Wealthier households are able to acquire brides for their sons earlier that poorer ones, and thus the sons of wealthy families can begin to bear children and to move toward establishing independent households sooner.

### The sexual division of labor

During much of the history of each household, the core of the household consists of a group of closely related men who cooperate under the direction of the man who is father and household head. The sexual division of labor among the Yomut places most of the economically productive labor in the hands of men. Men do virtually all of the shepherding. The only exception is that occasionally when lambs and kids are very young and are herded near the family yurt, girls as well as boys may do some of the shepherding. However, most of the shepherding involves tending to animals far from the family yurt and includes sleeping out at night with the animals away from the rest of the family. At the time of my field studies, there were still dangers from both wolves and thieves while shepherding. In the not too distant past, livestock raiding was common and further increased the dangers of shepherding. The agricultural labor of the *charwa* usually is also carried out far from the family home base. This also is considered male work. Among the *chomur* the site of agricultural work may be closer to the family residence, but care of flocks of sheep and goats is still carried our far from the family home.

The Yomut sexual division of labor represents a pattern that is widely associated with plow agriculture and pastoralism. Work away from the family residence is generally a male activity in such societies (Brown, 1970). This is part of a pattern in which women's work must be compatible with child care, and men must assume responsibility for heavy and dangerous work. In the period covered by the demographic data in the 1973–74 survey, the total fertility rate (TFR; the average number of children a woman bore if she lived through her reproductive years) was 7.5. Women ordinarily breastfed each child until another was born, and birth intervals averaged 33 months. This means that typically women in their 20s and 30s were either pregnant or breastfeeding most of the time, and in addition often were responsible for other children in addition to the most recently born. Such childcare responsibilities were not compatible with heavy or dangerous work away from the family's home base.

### Male labor, female labor, and wealth

The survey data from 1973–74 can be used to estimate the effect of the male labor resources of each household on the household's wealth. The survey data included a record of the wealth in land, livestock, and occasionally other items given to the household head at the time when the household became independent. It also includes the household's current wealth and the history of the household's membership over time, that is how many sons, daughters, wives, etc. were in the household throughout its history from the time of its founding to the time of the survey. Earlier data had suggested that households that had large male labor pools were better able to enhance their wealth over time (Irons, 1975, pp. 161–163). This earlier data had shown that men who had large numbers of sons were on average wealthier that those with few sons. The 1973–74 survey data allow a better test of the hypothesis that an increase in male labor tends to translate into an increase in wealth over time.

To compare wealth among households, the holdings of each household in livestock and land were converted into monetary values using the values of land and livestock in Iranian currency at the time of the survey (see Irons, 1980 for more details). The actual units used for the regression analysis below were thousands of *tomans* in 1974 prices. The male-labor-enhances-wealth hypothesis was tested with the survey data by regressing the value of each family's wealth at the time of the survey ($W_2$) on the following variables: wealth at the time the household was established ($W_1$), the number of male labor years the household had experienced ($L_m$), the number of female labor years ($L_f$), and the number of consumer years ($C$) between its establishment and the time of the survey. Each male member of the household was considered to add a male labor year to the household each year that he resided there after reaching age 15 years. Female labor years were calculated in same manner, and each year that any individual of any age resided in the household for a year they added a consumer year to the household's experience. The assumption that both men and women begin to contribute significant labor to the household at age 15 years was based on ethnographic observations (Irons, 1975, 1980). This analysis was done only for those households that had not yet had sons separate off, so that they represent households in the first stage of development in which wealth is amassed, sons begin to be married (which requires expenditure of a bridewealth) and the wealth that later allows sons to establish independent households is accumulated. The initial regression equation took the following form:

$$W_2 = 103.6 + 0.7 W_1 + 0.3 L_m - 0.06 L_f - 0.15 C \ (n = 384; r^2 = 0.55)$$

| Variable | $F$ |
|----------|---------|
| $W_1$ | 461.9** |
| $L_m$ | 12.1* |
| $L_f$ | 0.4 |
| $C$ | 14.7* |

*$p < 0.01$; **$p < 0.001$.

This equation reveals that a household's accumulation of wealth during this early part of its growth is influenced by the wealth it starts with, the male labor put into accumulating wealth over its lifetime and the number of consumers that must derive their livelihood from the household's income. Female labor, however, has no effect. (Note both males and females of all ages are counted as consumers.) As noted above, the developmental cycle can be seen as a process of accumulating wealth and then turning wealth into people. The above regression analysis suggests that the business of accumulating wealth is primarily a male activity and the business of turning wealth into people is primarily a female activity. Thus while women are absolutely essential to the process of development, their role consists of converting wealth into people, that is giving birth and rearing children. The fact that wives are obtained through the payment of a large bridewealth also helps to explain why adding female labor does not enhance wealth. Initially it has the effect of diminishing wealth.

My impression from a long period of observation of Yomut society suggested that the situation of a household most conducive to the accumulation of wealth was to have capital in both livestock and land and to have a large male labor pool for purposes of managing this capital. It also seems that the more wealth a household had the more each member would consume, that is the higher their standard of living would be. In order to test these intuitions, I regressed wealth at the time of the survey ($W_2$) onto wealth at the time of founding ($W_1$), the product of wealth at founding and male labor years ($W_1 \times L_m$), and the product of consumer years and wealth at founding ($C \times W_1$). Regressing $W_2$ on the product of $W_1$ and $L_m$ is meant to test for the hypothesized interaction of wealth and labor, and regressing onto the product of $W_1$ and $C$ is likewise intended to test for the interaction of wealth and individual rates of consumption. This analysis yielded the more satisfactory regression equation below, and confirmed the correctness of the intuitions about the interaction of male labor and capital and consumption and capital.

$$W_2 = 62.7 + 0.91 W_1 + 0.39(L_m \times W_1) - 0.13(C \times W_1) \ (n = 384; \ r_2 = 0.69)$$

| Variable | $F$ |
|---|---|
| $W_1$ | 693.7** |
| $L_m \times W_1$ | 45.0* |
| $C \times W_1$ | 70.1* |

*$p < 0.05$; **$p < 0.001$.

The fact that is most significant is that the best way to accumulate wealth is for a group of males to cooperate in managing capital, and, as in most traditional societies, people prefer to carry out important forms of cooperation with closely related individuals. Thus the Yomut habit of building households around closely related male kin, fathers and sons and brothers, makes economic sense. Since wealth is also then converted into reproduction for all the closely related men in the household, it also makes sense as an inclusive-fitness maximizing strategy. The political importance of patrilineal ties among men explained above also reinforce the advantage of cooperation among men closely related in the male line.

### Norms of interpersonal behavior within a household

While the Yomut are egalitarian in their assumptions about the relationships among households, they are decidedly not egalitarian concerning relationships within households. The head of a young household consisting of a nuclear family is the husband and father and the family's wealth is considered to be his wealth. The property a wife brings to her husband's household on marriage, her dowry, is an exception and is the property of the wife (later to be inherited by her children). However, for the majority of Yomut, this is a trivial amount of wealth. Usually it consists of jewelry, carpets, and other textile furnishings for a household, and some clothing. Among the very wealthy, it may also include livestock, but rarely a large number of animals. Wives are expected to be obedient to their husbands and the children of the household are expected to be obedient to their parents. While these norms of obedience may sound familiar as norms of traditional family life in North America, there is much more of an expectation of real obedience and deference than one is likely to encounter in even the most traditional of families in English-speaking North America. As sons reach adulthood they are expected to continue deferring to their father and to continue showing respect by avoiding casual conversation, joking, or smoking in his presence. To a degree, they are to continue to be seen much more than heard in his presence. Younger brothers are expected to show similar deference to older brothers. The daughters are similarly to show deference to their brothers, father and to their mother. Most of the activities of the household are segregated by sex. Women cluster together in the woman's part of a yurt or house, and talk among themselves. Men cluster together in the men's area so that interaction between closely related men and women is limited.

When a nuclear household grows into an extended household through the addition of a daughter-in-law, even stronger rules of deference come into play. Initially a daughter-in-law spends her days in her father-in-law's household, sitting behind a curtain that separates off a part of the yurt or house so that she can not be seen by either of her parents-in-law or any of her brothers-in-law who are older than her husband. She also is to stay out of the view of wives of her husband's older brothers. Individuals in the household who are younger than her husband, however, may interact with her. Thus daughters of the household who have not yet married out can join her behind the curtain where she busies herself with sewing, spinning, or other work that can be done in a confined space.

Over time, the restrictions on a daughter-in-law become less confining. After a short period the curtain is removed and the new daughter-in-law shows her deference to those who are senior to her by covering her face with her head cloth, avoiding talking to them, not eating with them, and not letting her mouth be seen when she is eating. Those who are senior to her include her father-in-law, her mother-in-law, her husband's older brothers and their wives. Nor in the presence of these individuals is she allowed to speak to her husband or to eat with him. As a woman has children her status rises somewhat within her husband's household, and sometimes her parents-in-law may invite her to discontinue covering her face in their presence. This is unusual, however, and occurs primarily when the daughter-in-law is a first cousin to her husband and a niece to her parents-in-law.

There are also rules of avoidance between a man and his parents-in-law and older brothers and sisters of his wife. These rules require him to avoid being in the same yurt or room in a house with them and to avoid talking to them. However, because they do not reside with their affines, these rules confine the activities of men much less than the rules of affine avoidance confine women.

These rules limiting contact and casual interaction between affines fall in the category of rules described in the anthropological literature as affine avoidance. Early in the history of anthropology, some structuralists attempted to explain these rules as ways of avoiding interacting with individuals when the norms of the society enjoin contradictory forms of behavior toward affines (Radcliffe-Brown and Forde, 1949). However, to me the norms as they applied to women had much more to do with simple, straightforward subordination. The ways a daughter-in-law is expected to behave toward her parents-in-law made sense to me as parallels to the ways I was expected to behave toward officers when I was a draftee in the US Army. Privates and non-commissioned officers did not speak to officers unless spoken to (other than formal greetings and salutes). They did not eat with them, and they did not relax in their presence. These norms reinforced the demand for unquestioning obedience. The Yomut themselves did not offer elaborate explanations of these rules. As with all other norms of behavior they tended to explain them in terms of respect, shame, or an obligation to do what their ancestors had done. According to Yomut friends, married-in daughters-in-law avoided senior members of the household to show respect, to express shame, or to fulfill an obligation to do what their ancestors had done. As it turns out, the US Army also did not offer explanations for their norms of avoidance that would impress a social scientist. They simply said this is the Army way, follow the rules or we will make you regret not following the rules. For Yomut men, the rules of affine avoidance were less of a burden since they did not live with their affines. For men, these rules had the effect of keeping sons-in-law from interfering in the running of the wives' natal households.

The subordinate position of women in the households of their husbands was accompanied by an assumption that they were of less importance than the other members of the household. When they were sick, for example, it was generally considered less urgent to seek medical attention. The best portions of an extended household's food were always placed on a tray for the head of the household and his sons to eat from. The household head's wife and her daughters would receive less desirable portions and sometimes would simply finish leftovers on the men's tray. Married-in daughters-in-law would eat separately receiving still less desirable portions. On occasion, Yomut friends did state that brothers were more important members of one's household than wives because they could not be replaced if they died. Wives could be replaced, although the bridewealth for a widower seeking a virgin bride would be higher than for a never-married man seeking a virgin bride. Several friends also noted that brothers were more "useful" than wives or sisters when one was caught up in a dispute of some sort. Brothers could threaten violence to back up one's claims in a dispute. Wives and sisters could not do this. Thus, the situations Yomut men faced in their daily lives tended to reinforce their lower valuation of women.

I suggest that the lower status of women in Yomut society and their lower importance by local standards during their reproductive years explains their higher death rates as revealed by the data below. Note that with equal amounts of medical care, equal opportunities to rest when ill, and equal access to food, in short equal social support, one would expect women to have lower death rates during these years as is the case in the majority of human populations.

### Male and female mortality rates: a reflection of status

The 1973–74 survey provides two bodies of data that allow estimates of death rates. The survey asked each household who in that domestic unit had died in the past 5 years and what the age, sex, and relationship to the household head was for each deceased individual. Cause of death was also recorded. Further, the survey included a census of each household so that both the number of deaths and the person-years at risk of death can be calculated for the sample population. These data can be used to estimate death rates for people of all ages. I refer to this body of data as the 5-year death register.

The survey also asked about all children born to women of the household including deceased wives of men of the household. The goal was to get complete fertility records and complete survivorship records for children born to both the women and the men of the household. The records of deaths among children born to these women allow estimation of death rates for the earlier years of life, that is, up to about age 20 years. I refer to this body of information as the child survivorship data.

Because these child survivorship data are based on a much larger sample, they provide better estimates of mortality and survivorship for the early portion of the life cycle. However, the sample of deaths in the later years of life in the child survivorship data is too small to allow for reliable estimates. In comparing the two bodies of data and the death rate estimates they yield, one also has to be aware that the death rates among the Yomut have been going down during the several decades from the end of World War II to the time of the survey (Irons, 1980, pp. 429–434). Thus the mortality estimates from the child survivorship data cover roughly the past 20 years, while the data in the 5-year death register cover only the past 5 years, a time period in which death rates were lower. However, for

Table 11.1. *Difference in Yomut male and female mortality rates for the first 15 years of life in the 5-year death register*

| Age interval (years) | Person-years at risk of death | Deaths | Death rate (deaths/year during 5-year interval) | Variance in sample mean of death rate |
|---|---|---|---|---|
| *Male mortality* | | | | |
| 0 | 489 | 61 | 0.1247 | 0.0002551 |
| 1 | 390 | 18 | 0.0462 | 0.0001183 |
| 2 | 376 | 6 | 0.0160 | 0.0000424 |
| 3 | 365 | 6 | 0.0164 | 0.0000450 |
| 4 | 359 | 3 | 0.0084 | 0.0000232 |
| 5–9 | 1465 | 5 | 0.0034 | 0.0000023 |
| 10–14 | 1323 | 4 | 0.0030 | 0.0000022 |
| | | | | |
| Age-standardized death rate | | | 0.2181 | |
| Variance in sample mean for age-standardized death rate | | | | 0.0004885 |
| | | | | |
| *Female mortality* | | | | |
| 0 | 472 | 57 | 0.1208 | 0.0002558 |
| 1 | 385 | 23 | 0.0597 | 0.0001551 |
| 2 | 381 | 12 | 0.0315 | 0.0000826 |
| 3 | 333 | 2 | 0.0060 | 0.0000180 |
| 4 | 302 | 2 | 0.0066 | 0.0000219 |
| 5–9 | 1357 | 3 | 0.0022 | 0.0000016 |
| 10–14 | 1155 | 3 | 0.0026 | 0.0000022 |
| | | | | |
| Age-standardized death rate | | | 0.2294 | |
| Variance in sample mean for age-standardized death rate | | | | 0.0005372 |

Z-score for difference in male and female age-standardized death rates = 0.35; $p = 0.36$.

purposes of comparing male and female mortality this is not a problem as long as comparisons are made among estimates derived from the same body of data.

### Mortality estimates derived from the five-year death register

The estimates of death rates derived from these data for men and women are presented in Tables 11.1 through 11.4. Table 11.1 presents the estimated death rates for the first 15 years of life (ages 0–14). Table 11.2 presents the estimates for the age range 15 to 49 years which corresponds to the reproductive years for women, and Table 11.3 presents the estimates for the age range 50 to 74 years. The number of individuals in the sample

Table 11.2. *Difference in Yomut male and female mortality rates for ages 15–49 years in the 5-year death register*

| Age interval (years) | Person-years at risk of death | Deaths | Death rate (deaths/year during 5-year interval) | Variance in sample mean of death rate |
|---|---|---|---|---|
| *Male mortality* | | | | |
| 15–19 | 1050 | 2 | 0.0019 | 0.0000018 |
| 20–24 | 848 | 0 | 0 | 0 |
| 25–29 | 633 | 0 | 0 | 0 |
| 30–34 | 538 | 0 | 0 | 0 |
| 35–39 | 551 | 1 | 0.0018 | 0.0000032 |
| 40–44 | 399 | 2 | 0.0050 | 0.0000125 |
| 45–49 | 377 | 5 | 0.0133 | 0.0000351 |
| Age-standardized death rate | | | 0.0220 | |
| Variance in sample mean for age-standardized death rate | | | | 0.0000526 |
| *Female mortality* | | | | |
| 15–19 | 933 | 6 | 0.0064 | 0.0000067 |
| 20–24 | 895 | 3 | 0.0034 | 0.0000037 |
| 25–29 | 607 | 1 | 0.0016 | 0.0000026 |
| 30–34 | 405 | 4 | 0.0099 | 0.0000240 |
| 35–39 | 423 | 2 | 0.0047 | 0.0000290 |
| 40–44 | 411 | 5 | 0.0122 | 0.0000290 |
| 45–49 | 394 | 3 | 0.0076 | 0.0000188 |
| Age-standardized death rate | | | 0.0458 | |
| Variance in sample mean for age-standardized death rate | | | | 0.0001138 |

$Z$-score for difference in male and female age-standardized death rates $= 1.85$; $p = 0.03$.

above age 74 years was too small to allow good estimates. Table 11.4 presents survivorship rates for men and women in the same intervals derived from the 5-year death register.

The tests for statistical significance used here were devised by Clifford Clogg, formerly of the Pennsylvania State University Department of Sociology. These tests are based on the assumption that the death rates have a Poisson distribution, and therefore that the variances in the sample mean for estimates of the death rate will consist of the death rate divided by the sample size. An age-standardized death rate was then calculated by summing the death rates for each interval, and a variance in the sample mean for this standardized rate was calculated by summing the variance for each age interval. This procedure corresponds to using weightings of one for each age interval as suggested by Kitagawa (1964). Assuming the estimates

Table 11.3. *Difference in Yomut male and female mortality rates for ages 50–74 years in the 5-year death register*

| Age interval (years) | Person-years at risk of death | Deaths | Death rate (deaths/year during 5-year interval) | Variance in sample mean of death rate |
|---|---|---|---|---|
| *Male mortality* | | | | |
| 50–54 | 269 | 4 | 0.0149 | 0.0000552 |
| 55–59 | 196 | 3 | 0.0153 | 0.0000780 |
| 60–64 | 174 | 5 | 0.0287 | 0.0001651 |
| 65–69 | 102 | 7 | 0.0686 | 0.0006728 |
| 70–74 | 86 | 5 | 0.0581 | 0.0006760 |
| Age-standardized death rate | | | 0.1856 | |
| Variance in sample mean for age-standardized death rate | | | | 0.0016471 |
| *Female mortality* | | | | |
| 50–54 | 178 | 2 | 0.0112 | 0.0000631 |
| 55–59 | 172 | 2 | 0.0116 | 0.0000676 |
| 60–64 | 125 | 6 | 0.0480 | 0.0003840 |
| 65–69 | 78 | 1 | 0.0128 | 0.0001643 |
| 70–74 | 25 | 2 | 0.0800 | 0.0032000 |
| Age-standardized death rate | | | 0.1636 | |
| Variance in sample mean for age-standardized death rate | | | | 0.0038790 |

$Z$-score for difference in male and female age-standardized death rates $= 0.16$; $p = 0.44$.

Table 11.4. *Survivorship data: percent surviving various age intervals based on the 5-year death register*

| Age interval (years) | Male | Female | $Z$-scores | $p$-value |
|---|---|---|---|---|
| 0–14 | 77.6 | 77.2 | 0.35 | 0.36 |
| 15–49 | 89.5 | 79.5 | 1.85 | 0.03 |
| 50–74 | 38.6 | 43.1 | 0.16 | 0.44 |

$Z$-scores and $p$-values are for the differences in male and female age-standardized death rates.

of the variance in sample mean for the standardized death rates approximate a normal distribution, it is possible to calculate the $Z$-scores and $p$-values presented in Tables 11.1 through 11.4. These statistical techniques were explained in an earlier publication (Irons, 1980).

The $p$-values indicate that there is no statistically significant difference in the death rates of Yomut men and women in the age intervals 0 through 14 and 50 through 74 years. However, women have a statistically significant higher death rate than men in this sample for the female reproductive years, 15 through 49 years.

This higher female death rate could be interpreted as a result of the burden of childbearing. The survey data record 595 births in the sample population in the past 5 years and only four deaths attributed to childbirth. The cause of death was recorded in the survey interviews for those deaths occurring in the past 5 years and what was recorded was what the surviving members of the deceased's household stated in the survey interview as the cause of death. Assuming these perceptions of the cause of death are accurate, the risk of death in childbirth was roughly 7 in 1000. Note that the Yomut, with very rare exceptions, give birth at home with only the assistance of traditional midwives and other female members of their household. If there are serious difficulties, attempts may be made to transport the woman in trouble to a free government clinic or to a hospital in the city of Gonbad-e Kavus, but often such aid is sought too late. Also poorer families and those in remoter areas are much less likely to seek the aid of anyone other than a traditional healer in such cases. At the time of the survey effective modern medical care for childbirth, or other situations where it may have been required, was unusual and sporadic. The low risk of death in childbirth was surprising to the researchers doing the survey in light of this limited use of modern medical care. On the other hand, the childbearing years are years of low death rate in general. Only 24 deaths were recorded in the 5 years before the survey for women in this age range and of them four were due to childbirth. Thus, although the risk of death from childbirth was low for this population during the 5 years preceding the survey, it still ranks as one of the major risks within this age range.

On the other hand, there were no cases of death from accident for women in the childbearing years, but for men in the same age range there were three deaths by accident out of the total of 11 deaths in the same age range (15–49 years). Thus it seems that for the Yomut as for most human populations the risk of death for each sex comes from different sources. It also needs to be noted that the sample of deaths for which causes were recorded is small and the number of cases where the response was unknown (or at least not stated) was high; these data are merely suggestive as to the relative frequency of causes of death.

These empirical survivorship and mortality rates presented in Tables 11.1 through 11.4 can also be used to construct a period life table for the

Table 11.5. *Life expectancy at various ages estimated from the*
*5-year death register*

| Age | Male | Female |
|-----|------|--------|
| At birth | 53.1 | 49.5 |
| 15 years | 52.9 | 48.6 |
| 50 years | 20.8 | 21.1 |

5-year time period covered by the 5-year death register. An abbreviated
form of this life table is presented in Table 11.5.[1]

### Mortality estimates derived from child survivorship data

As noted above, these data allow estimates of mortality from a larger
sample for the first 20 years of life. This body of data includes a record of
981 deaths for the first 20 years of life while the 5-year death register
analyzed above includes only 205 deaths for the same age range. However,
as noted above, these data do not provide an adequate sample for the later
years of life.

---

[1] The life table presented here in abbreviated form was computed using the techniques described
in Pressat (1972) with some modifications. Pressat's techniques are designed primarily for
recent European data. Determining the number of years that very old individuals will live is
difficult in any population because of limited data. Pressat recommends assuming that all
individuals who reach age 90 will live for 2.5 additional years. He also notes that assuming
different numbers of years for people near the end of life has little effect on the final life
expectancies computed because they represent a very small portion of the person-years lived by
the entire population. I modified this assumption by assuming that all persons reaching age 80
would live an additional 2.5 years. There were three individuals in the sample population of
4096 who were 80 years or older, and the 5-year death register included only three cases of
individuals dying in the 80 or older category. In light of this an assumption of 2.5 years of life
on average after an individual's eightieth birthday seemed reasonable. All of the other data
used to compute death rates, percentage of individuals surviving specific age intervals, and life
expectancies were purely empirical. They were taken out of the survey data without any
smoothing or modifications based on assumptions about inaccuracies in the data recorded. This
includes empirical death rates for the age categories 75–79 years which were not included in
Tables 11.3 and 11.4 because the sample was small. These death rates were 0.097 for males and
0.25 for females. The difference is most reasonably interpreted as sampling error. It could be
argued that I should have moved from empirical data to assumptions at age 75 years. However,
the important point made by Pressat is that, while it is necessary to use either small samples or
assumptions to complete the calculation of a life table, it makes little difference what is done,
since different assumptions or uses of small samples of data have little effect on the final
numbers. The overwhelming majority of individuals die before reaching the ages at which the
empirical data peter out, so to speak.

Table 11.6. *Difference in Yomut male and female mortality rates for the first 2 years of life estimated from the child survivorship files*

| Age interval (years) | Person-years at risk of death | Deaths | Death rate (deaths/year during 5-year interval) | Variance in sample mean of death rate |
|---|---|---|---|---|
| *Male mortality* | | | | |
| 0 | 1934 | 285 | 0.147 | 0.000076 |
| 1 | 1525 | 107 | 0.070 | 0.000046 |
| Age-standardized death rate | | | 0.217 | |
| Variance in sample mean for age-standardized death rate | | | | 0.000122 |
| *Female mortality* | | | | |
| 0 | 1731 | 225 | 0.130 | 0.000075 |
| 1 | 1408 | 84 | 0.060 | 0.000042 |
| Age-standardized death rate | | | 0.190 | |
| Variance in sample mean for age-standardized death rate | | | | 0.000117 |

*Note:* Z-score for difference in male and female age-standardized death rates = 1.75; $p = 0.04$.

Also as noted above, one needs to keep in mind that these data are from a period in which the death rate was higher on average than in the period covered by the 5-year death register. Survivorship data based on the 5-year death register show 77.6% of males and 77.2% of females surviving from birth to their fifteenth birthday. Survivorship data drawn from the child survivorship files show 69% of males and 74% of females surviving the same interval. This difference should not, however, bias the results as long a one compares male and female death rates within the same body of data.

Analysis of the data in the child survivorship files yields somewhat different results, but not different enough, in my opinion, to alter the conclusions. These data show a significantly higher death rate for males in the first 2 years of life, and no significant difference in male and female death rates for ages 2 through 19 years. These data are summarized in Tables 11.6 through 11.8. I attribute the difference between the results from the different bodies of data to sample size, and assume the large sample yields the more accurate estimates. The first 2 years of life are years of great vulnerability and apparently the greater solicitude for sons over daughters is not sufficient in these early years to reverse the biologically based pattern of higher male mortality. Nevertheless the advantage females enjoy in this early age interval is smaller than the advantage enjoyed by males in the age interval 15–49 years. The data show a difference of

Table 11.7. *Difference in Yomut male and female mortality rates for ages 2 through 19 years estimated from the child survivorship files*

| Age interval (years) | Person-years at risk of death | Deaths | Death rate (deaths/year during 5-year interval) | Variance in sample mean of death rate |
|---|---|---|---|---|
| *Male mortality* | | | | |
| 2 | 1343 | 60 | 0.045 | 0.000033 |
| 3 | 1211 | 36 | 0.030 | 0.000024 |
| 4 | 1110 | 13 | 0.012 | 0.000010 |
| 5–9 | 4448 | 31 | 0.007 | 0.000002 |
| 10–14 | 3175 | 8 | 0.003 | 0.000001 |
| 15–19 | 2065 | 4 | 0.002 | 0.000001 |
| Age-standardized death rate | | | 0.099 | |
| Variance in sample mean for age-standardized death rate | | | | 0.000071 |
| *Female mortality* | | | | |
| 2 | 1273 | 61 | 0.048 | 0.000038 |
| 3 | 1127 | 28 | 0.025 | 0.000022 |
| 4 | 1018 | 6 | 0.006 | 0.000006 |
| 5–9 | 4191 | 32 | 0.008 | 0.000002 |
| 10–14 | 2898 | 5 | 0.002 | 0.000001 |
| 15–19 | 1870 | 9 | 0.005 | 0.000002 |
| Age-standardized death rate | | | 0.094 | |
| Variance in sample mean for age-standardized death rate | | | | 0.000071 |

$Z$-score for difference in male and female age-standardized death rates = 0.26; $p = 0.40$.

Table 11.8. *Survivorship data from the child survivorship files: percent surviving birth to second birthday and second to twentieth birthday*

| Age interval (years) | Male | Female | $Z$-score | $p$-value |
|---|---|---|---|---|
| 0–1 | 76.2 | 81.8 | 1.75 | 0.04 |
| 2–19 | 89.7 | 88.3 | 0.261 | 0.40 |

$Z$-scores and $p$-values are for the differences in male and female age-standardized death rates.

5.6% favoring females over males in the age interval from birth to the second birthday (see Table 11.8). The death register sample shows a difference of 10% advantage favoring males in the age interval corresponding to the female reproductive years (see Table 11.4).

### Conclusion regarding mortality

What is clear is that the Yomut have higher death rates for women in the reproductive years and approximately the same death rates for males and females during most of the rest of the life cycle. The exception to the latter pattern is the higher death rate for males in the first 2 years of life. The advantage enjoyed by females in the first 2 years of life is smaller than the advantage enjoyed by males in the middle of the life cycle. The overall effect of these death rates is a shorter life expectancy at birth for women and a shorter life expectancy at various later ages as well (see Table 11.5). When these data are compared with data from other populations, this condition stands out as unusual.

The World Almanac and Book of Facts 2000 lists life expectancies for 191 nations. Only seven of these (Afghanistan, Bangladesh, Bhutan, Malawi, Namibia, Nepal, and Niger) have higher life expectancies for men than women, all but six have a male life expectancy that is less than a year longer than female life expectancy. Only Afghanistan has a difference of a year with a male life expectancy of 47.82 and a female life expectancy of 46.82. Data of this type are different in many ways from the data presented on the Yomut. Entire nations are more heterogeneous in terms of culture and healthcare. It would be more instructive to compare the Yomut with other small culturally homogeneous populations. Other useful comparisons would be with the !Kung, the Ache, and the Yanomamo. These populations are all found in context of much simpler technologies and greater isolation from modern urban society than the Yomut. Nevertheless they all share with the Yomut limited access to modern medical care, and virtually no use of modern contraceptives. However, in cultural tradition these populations are all very different from the Yomut. Part of this difference appears to be less of a tendency to undervalue women. The Ache (Hill and Hurtado, 1996), under traditional conditions, have a slightly higher life expectancy for men than women. Ache men living a traditional foraging lifestyle in the forest can expect to live 37.8 years. Ache women living the same lifestyle can expect to live 37.1 years. However, during the reproductive years Ache women living the traditional way have lower death rates than men. The male advantage is greater for Ache living on reservations. For this group men have a life expectancy of 50.4 years and women have a life expectancy of 45.6 years. These life expectancies are much closer to the ones presented here for the Yomut. In contrast, the !Kung data (Howell, 1979) and the Yanomamo data (Melancon, 1982; Early and Peters, 1990) both show the more general pattern of women having a longer life expectancy and lower death rates than men.

Theoretical considerations also suggest that in general, one should expect male death rates to be higher than female death rates at all ages in any species or population that has a history of polygynous breeding (Trivers, 1972; Alexander, 1979; Daly and Wilson, 1983). The available evidence points to a history of mildly polygynous breeding in human populations, and overall the published mortality data for different human populations show strong empirical support for this tendency. Human male development and physiology are characterized by greater vulnerability to a number of hazards, and, on both theoretical and empirical grounds, I believe that with something approaching equal levels of stress and social support, males will always have higher death rates than females at all ages. In the case of the Yomut, females across their lifetimes receive less social support. In the years before the reproductive portion of their lifetimes, women have lower death rates only for the first 2 years of life. From 2 through 19 years they have approximately the same death rates as males. This, I suggest, is a result of less social support for females. During the reproductive years the situation for Yomut women is even less desirable and, despite an overall physiological advantage, they experience higher death rates. Again following their reproductive years, male and female death rates are not significantly different according to the data analyzed here.

The differences between male and female mortality are not only influenced by a greater male vulnerability built into their physiology and development (Trivers, 1972; Alexander, 1979; Daly and Wilson, 1983), but by different degrees of support from close kin. The latter in the form of differential solicitude of parents for sons versus daughters has been extensively documented (Low, 2000). However, differential solicitude goes beyond the parent–child relationship. Most human beings spend their adult years co-residing and closely interacting with other members of kin-based domestic groups. The extent to which various adults assist rather than exploit members of their domestic groups varies from one society to another. In the case of the Yomut, close support is maintained among fathers and sons and between brothers (cf. Irons, 1979). For men these close kin are the primary source of support and male agnatic kin ties are given the highest value. Close female kin, sisters and daughters, are in effect sacrificed to the need to hold together a closely cooperating group of male agnates. Sisters and daughters move at marriage to assume a subordinate position in the households of their fathers- and mothers-in-law. As wives and daughters-in-law, they are incorporated as subordinates and outsiders into the households of their husbands. Only long after marriage does a woman regain a status similar to that of an unmarried daughter in her parents' household. This return to a higher status is fully achieved when a woman assumes the position of the female head of the female portion of her husband's independent household. Roughly the years of greatest stress and least support correspond to the reproductive years when a woman is a subordinate member of her parents-in-law's household. These years of least support and greatest stress are reflected, I am convinced, in the mortality rates presented in this chapter.

### Summary and conclusions

This chapter presents a range of data concerning the family organization of the Yomut Turkmen. The data supports several inferences:

1. Yomut domestic groups are built around close kinship ties between men. Yomut women's life histories need to be accommodated to the need of their husbands to maintain cooperation with their closest male kin.

2. This family organization has economic advantages and in the recent past had political advantages for the men involved. To some degree these political advantages are still present.

3. The integrity of the male core of Yomut domestic groups is maintained by subordinating the interests of married-in women to the interest of the male core of the group. This entails serious restriction on the activities of women during their reproductive years and a situation of stress and limited social support. This exacerbates a general tendency in Yomut society to value men more than women and to be generally more solicitous to the needs of men than women.

4. As a consequence of the greater concern for the welfare of men over that of women in Yomut society, women experience a lower life expectancy. This occurs despite the fact that sexual selection has favored male developmental and physiological traits that make them more competitive but also more vulnerable to a wide range of risks with consequent higher death rates under conditions of equal social support.

### Acknowledgments

The collection, coding, and analysis of this data was supported by the National Science Foundation Program in Cultural Anthropology, the Ford Foundations Program in Population Research and by the H. F. Guggenheim Foundation. The sampling procedures used in Iran were designed by Alan Ross of The Johns Hopkins University School of Public Health, Department of Population Dynamics. In 1976–77, Clifford Clogg then of the Sociology Department of The Pennsylvania State University served as a statistical consultant, and the tests of statistical significance used for demographic rates were developed by him. Analyses of the 1973–74 survey data have appeared in earlier publications (Irons, 1979, 1980, 1986, 1994, 2000).

### References

Alexander, R.D. (1979). *Darwinism and Human Affairs*. Seattle: University of Washington Press.

Barthold, V.V. (1962). *A History of the Turkmen People*. (Translated by V. and T. Minorsky.) Leiden: E. J. Brill.

Brown, J.K. (1970). A Note on the division of labor by sex. *American Anthropologist* **72**, 1073–1078.

Daly, M. and Wilson, M. (1983). *Sex, Evolution, and Behavior*, 2nd edn. Boston: Willard Grant Press.

Early, J. and Peters, J. (1990). *The Populations Dynamics of the Mucajai Yanomamo*. New York: Academic Press.

Hill, K. and Hurtado, M. (1996). *Ache Life History: The Ecology and Demography of a Foraging People*. New York: Aldine De Gruyter.

Hockett, C.F. (1958). *A Course in Modern Linguistics*. New York: Macmillan.

Howell, N. (1979). *Demography of the Dobe !Kung*. New York: Academic Press.

Irons, W. (1971). Variation in political stratification among the Yomut Turkmen. *Anthropological Quarterly* **44**, 143–156.

Irons, W. (1974). Nomadism as a political adaptation: the case of the Yomut Turkmen. *American Ethnologist* **1**, 635–658.

Irons, W. (1975). *The Yomut Turkmen: A Study of Social Organization Among a Central Asian Turkic Speaking Population*, Anthropological Paper Number 58. Ann Arbor, MI: Museum of Anthropology, University of Michigan.

Irons, W. (1979). Political stratification among pastoral nomads. In *Production Pastorale et Société*. Equip pour Anthropologie et Ecologie des Société Pastorale, ed. Maison des Sciences de l'Homme, pp. 361–374. New Rochelle: Cambridge University Press and Paris: Maison des Sciences de l'Homme.

Irons, W. (1980). Is Yomut social behavior adaptive? In *Sociobiology: Beyond Nature/Nurture?* ed. G.W. Barlow and J. Silverberg, pp. 417–473. American Association for the Advancement of Science Series on Science: The State of the Art. Boulder, CO: Westview Press.

Irons, W. (1986). Yomut family organization and inclusive fitness. In *Proceedings of the International Meetings on "Variability and Behavioral Evolution,"* pp. 227–236. Rome: Accademia Nazionale dei Linei.

Irons, W. (1994). Why are the Yomut not more stratified? In *Pastoralists at the*

*Periphery: Herders in a Capitalist World*, ed. C. Chang and H.A. Koster, pp. 275–296. Tucson, AZ: The University of Arizona Press.

Irons, W. (2000). Why do the Yomut raise more sons than daughters? In *Adaptation and Human Behavior: An Anthropological Perspective*, ed. L. Cronk, N.A. Chagnon and W. Irons. Hawthorn, NY: Aldine De Gruyter.

Low, B.S. (2000). *Why Sex Matters: A Darwinian Look at Human Behavior.* Princeton, NJ: Princeton University Press.

Melancon, T. (1982). Marriage and Reproduction among the Yanomamo Indians of Venezuela. PhD dissertation, The Pennsylvania State University.

Pressat, R. (1972). *Demographic Analysis.* Translated by Judah Matras. New York: Aldine.

Radcliffe-Brown, A.R. and Forde, D., ed. (1949). *African Systems of Kinship and Marriage.* Oxford: Oxford University Press.

Trivers, R.L. (1972). Parental investment and sexual selection. In *Sexual Selection and the Descent of Man*, ed. B. Campbell, pp. 136–179. Chicago: Aldine.

*World Almanac and Book of Facts 2000* (2000). Mahwah, NJ: World Almanac Books.

# 12 *Pastoralism and the evolution of lactase persistence*

CLARE HOLDEN AND RUTH MACE

## Introduction

In most parts of the world, the majority of adults are unable to digest lactose, or "milk sugar", the principle carbohydrate found in milk. Infants can digest lactose, but in most people, lactose digestion capacity declines sharply after infancy. This is a standard mammalian developmental pattern. In contrast, in some populations, especially Northern Europeans and pastoralists in Africa and the Middle East, the majority of adults retain the ability to digest lactose into adulthood. The physiological cause of high lactose digestion capacity in adulthood is the retention of high levels of the enzyme lactase, used to digest lactose, in the small intestine beyond infancy (Flatz, 1987; Durham, 1991). This is also known as lactase persistence. Variation in lactose digestion capacity in adulthood is associated with a single-locus genetic polymorphism. Lactase persistence is inherited as a dominant trait (Sahi *et al.*, 1973; Ransome-Kuti *et al.*, 1975; Johnson *et al.*, 1977; Metneki *et al.*, 1984).

The aim of this analysis was to test three hypotheses for the evolution of lactase persistence. The first hypothesis is that lactase persistence is a genetic adaptation to drinking milk from domestic livestock (Simoons, 1969, 1970a, 1978; McCracken, 1971; Flatz, 1987). This hypothesis implies coevolution between genes and culture, because pastoralism, a cultural trait, is hypothesized to have caused selection for lactase persistence, a genetic trait (Aoki, 1986; Feldman and Cavalli-Sforza, 1989; Durham, 1991). It is hypothesized that the capacity to digest lactose has a selective advantage to adults in pastoralist populations, because lactase-persistent adults are able to derive a nutritional benefit from lactose in milk, which is not available to non-persistent individuals. Lactose accounts for about 30% of the calorific content of cows' milk (Mustapha *et al.*, 1997). Non-lactase persistent individuals may also suffer from symptoms of lactose intolerance when they consume fresh milk, including abdominal discomfort, flatulence and diarrhea. It has been suggested that because of these

280

symptoms milk could be nutritionally detrimental to non-lactase persistent individuals, although this suggestion is controversial (see Scrimshaw and Murray, 1988). Populations which keep livestock but do not milk them, for example in China, Southeast Asia and parts of sub-Saharan Africa (Murdock, 1967; Simoons, 1970*b*) may not have been selected for lactase persistence (Simoons, 1979).

Milk can also be consumed in a processed form, such as cheese or soured milk, which has a reduced lactose content. Two selective pressures specifically for drinking fresh milk, with a high lactose content, have been proposed. In high-latitude environments, where sunshine is limited, humans are at risk of vitamin D deficiency and rickets. Flatz and Rotthauwe (1973) hypothesised that lactose in fresh milk, like vitamin D, promotes the uptake of calcium, also present in milk. This hypothesis is supported by the high frequencies of lactase persistence found in Northern European populations (Sahi, 1994; see Table 12.1). This hypothesis was used by Durham (1991) to account for the contrast between Northern and Southern Europeans – the latter traditionally processed milk into cheese, and remained predominantly non-lactase persistent, despite the antiquity of domesticated livestock in the Mediterranean. Experimental evidence for the role of lactose as a promoter of calcium uptake, in lactase-persistent individuals, is reviewed by Mustapha *et al.* (1997).

The third hypothesis is that drinking fresh milk may be a valuable source of water in highly arid environments, helping to maintain the body's electrolyte balance (Cook and Al-Torki, 1975; Cook, 1978). Cook and Al-Torki hypothesized that in arid environments, diarrhea and consequent water depletion in non-lactase persistent pastoralists would have caused selection against those individuals. This hypothesis is supported by the high frequencies of lactase-persistent adults found in pastoralist groups in hot, arid areas in the Middle East and North Africa, including the Bedouin, Tuareg and Fulani (Table 12.1).

One way to test these hypotheses statistically would be to test for an association between the hypothesized independent variables (pastoralism, solar intensity and aridity) and lactase persistence across populations. However, human populations are ancestrally related, so should not be treated as independent data points in a comparative analysis. In anthropology this is known as Galton's problem. Populations that share a common ancestor have many similarities, both cultural and genetic, which they inherited from their common ancestor (Mace and Pagel, 1994; Guglielmino *et al.*, 1995; Holden and Mace, 1999; Mace and Holden, 1999). These inherited similarities can produce statistical associations among traits across populations, which are not evidence of a functional

Table 12.1. *Lactase persistence in 62 populations worldwide*

| Population | Dry months | Solar radiation | Pastoralism (%) | Number of individuals tested | Lactase persistence (%) |
|---|---|---|---|---|---|
| Apache Nh17 | 12 | 162 | 0 | 22 | 0 |
| Australian Aborigines[a] | 12 | 167 | 0 | 45 | 16 |
| Baggara (Habbania) Cb13 | 10 | 174 | 50.5 | 19 | 47 |
| Baggara (Messiria) Cb15 | 10 | 174 | 70.5 | 20 | 40 |
| Bedouin (Saudi) Cj5 | 12 | 192 | 93 | 35 | 83 |
| Bedouin (Jordanian) Cj2 | 10 | 162 | 80.5 | 162 | 76 |
| Beja (Bisharin) Ca5 | 12 | 192 | 80.5 | 22 | 86 |
| Beja (Amarar) Ca35 | 12 | 192 | 93 | 82 | 87 |
| Beja (Beni Amir) Ca36 | 12 | 174 | 80.5 | 40 | 87 |
| Beja (Haddendoa) Ca43 | 12 | 174 | 60.6 | 137 | 80 |
| Chippewa Na36 | 7 | 124 | 0 | 33 | 3 |
| Czechs Ch3 | 9 | 124 | 30.5 | 200 | 87 |
| Dinka Aj11 | 5 | 144 | 50.5 | 213 | 24 |
| Egyptians Cd2 | 12 | 192 | 30.5 | 742 | 36 |
| Fijians[b] Ih4 | 0 | 167 | 0 | 12 | 0 |
| Fulani (pastoralist) Cb8 | 7 | 174 | 80.5 | 9 | 78 |
| Fulani (sedentary) Cb22 | 7 | 174 | 40.5 | 24 | 29 |
| Ganda Ad7 | 0 | 144 | 10.5 | 27 | 4 |
| Greeks Ce7 | 8 | 162 | 30.5 | 800 | 48 |
| Greenland Inuit Na25 | 8 | 79 | 0 | 119 | 15 |
| Hausa Cb26 | 7 | 174 | 30.5 | 17 | 24 |
| Hazara Tajiki Ea3 | 10 | 162 | 50.5 | 79 | 18 |
| Herero Ab2 | 9 | 178 | 60.5 | 37 | 3 |
| Hopi Nh18 | 10 | 162 | 10.5 | 21 | 0 |
| Hungarian Ch8 | 7 | 124 | 20.5 | 535 | 63 |
| Hutu Ae10 | 2 | 145 | 30.5 | 51 | 49 |
| Igbo Af10 | 4 | 144 | 10.5 | 16 | 19 |
| Iranian Ie9 | 12 | 162 | 30.5 | 40 | 17 (> 12 years) |
| Irish Cg3 | 0 | 94 | 40.5 | 50 | 96 |
| Italians (South) Ce5 | 4 | 124 | 10.5 | 197 | 33 |
| Japanese Ed5 | 1 | 162 | 10.5 | 66 | 19 |
| Javanese (Indonesia) Ib2 | 1 | 144 | 20.5 | 53 | 9 |
| Jordanians Cj6 | 10 | 162 | 30.5 | 204 | 25 |
| Lebanese Cj7 | 5 | 162 | 20.5 | 225 | 21 |
| Mongols Eb7 | 10 | 124 | 80.5 | 198 | 12 |
| Nama Aa3 | 9 | 178 | 50.5 | 18 | 50 |
| Northern Chinese (Han)[c] | 8 | 162 | 20.5 | 314 | 12 |
| Nubians (Midobi) Cb11 | 12 | 174 | 93 | 21 | 33 |
| Nuer Aj3 | 5 | 144 | 50.5 | 23 | 22 |
| Papago Ni2 | 10 | 162 | 0 | 14 | 7 |
| Pathans/Pushtu Ea2 | 10 | 162 | 30.5 | 86 | 35 |
| Pima Ni6 | 10 | 162 | 0 | 62 | 5 (> 4 years) |
| Punjabis Ea13 | 5 | 162 | 20.5 | 384 | 44 |
| Russians Ch11 | 6 | 94 | 30.5 | 103 | 43 |
| Sámi Cg4 | 8 | 79 | 60.5 | 519 | 59 |

Table 12.1 (*cont.*).

| Population | Dry months | Solar radiation | Pastoralism (%) | Number of individuals tested | Lactase persistence (%) |
|---|---|---|---|---|---|
| San (!Kung and #hua)[d] | 9 | 178 | 0 | 65 | 5 |
| Shilluk Ai6 | 10 | 144 | 20.5 | 8 | 37 |
| Sindhi Ea1 | 11 | 192 | 30.5 | 45 | 58 |
| Sinhalese[e] Eh6 | 0 | 144 | 30.5 | 158 | 27 |
| Sotho Ab8 | 5 | 178 | 30.5 | 23 | 35 |
| Spanish Ce6 | 5 | 162 | 30.5 | 265 | 85 |
| Swazi Ab2 | 4 | 192 | 20.5 | 12 | 25 |
| Tamils Eg2 | 0 | 144 | 20.5 | 31 | 29 |
| Thai Ej9 | 4 | 174 | 10.5 | 339 | 2 |
| Tswana Ab13 | 6 | 178 | 40.5 | 24 | 17 |
| Tuareg (Aulliminden) Cc8 | 8 | 144 | 60.5 | 118 | 87 |
| Tunisians Cd16 | 8 | 162 | 20.5 | 43 | 17 |
| Turks Ci5 | 6 | 162 | 40.5 | 470 | 29 |
| Tutsi Ae10 | 2 | 145 | 40.5 | 59 | 93 |
| Xhosa Ab11 | 3 | 178 | 30.5 | 17 | 18 |
| Yoruba Af6 | 4 | 144 | 10.5 | 100 | 9 |
| Zulu Ab12 | 3 | 178 | 40.5 | 32 | 19 |

*Note:* The name of each population is followed by its *Ethnographic Atlas* code. The following variables are shown: number of months per annum with <50 ml rainfall ("dry months"); global solar radiation (kcal/cm per year); pastoralism as a proportion of total subsistence ("Pastoralism %"); total number of individuals tested in each population; and frequency of lactase persistence in each population. Full references for the lactase persistence data are found in Holden and Mace (1997).
[a] *Ethnographic Atlas* not used. Subsistence practices and longitude and latitude taken from Brand *et al.* (1983).
[b] 6–15% Dependence on livestock, but pig-based livestock economy.
[c] Shantung Chinese, Murdock ref. Ed10, Cluster 163, used.
[d] Nyae Nyae !Kung, *Ethnographic Atlas* ref. Aa1, used.
[e] Longitude and latitude of Sri Lanka used, instead of *Ethnographic Atlas* coordinates.

relationship between those traits. Treating populations as if they were independent in a comparative analysis tends to inflate the statistical significance attached to associations between traits, and can also obscure real, recent correlated evolutionary change among variables against a background of ancestral dissimilarity (Harvey and Pagel, 1991).

To control for non-independence among populations, a phylogenetic comparative method was used here (Mace and Pagel, 1994). A genetic distance tree was used as a model of past relationships among populations, or phylogeny. In phylogenetic comparative methods, variables are mapped onto the tree. Character states at internal nodes on the tree (ancestral character states) are estimated from character states at the tips of the tree (in living populations in the sample). Mapping variables onto a population phylogeny enables one to identify independent instances of evolutionary change, because character change along any branch on a tree is independent of change on another branch. One can then test for evidence of correlated evolutionary change in variables, along the branches of the tree. The problem of non-independence among populations is thereby controlled for.

The phylogenetic comparative methods used here were developed in evolutionary biology, to control for non-independence among species. Applying these methods to human populations raises some interesting questions, relating to human population structure and to the transmission of traits between populations. We define populations as groups of individuals within which most reproduction occurs. Across populations, gene flow and cultural diffusion are possible, in contrast to species, among which there is no gene flow (Mayr, 1963). It is of interest to ask how important the "horizontal" diffusion of biocultural traits is, relative to "vertical" inheritance. Horizontal diffusion is defined as the transmission of biocultural traits between neighboring populations. Vertical inheritance is defined as transmission from "mother" to "daughter" populations. We tested whether the biocultural variables in the sample were associated with phylogeny, indicating vertical transmission, and/or geographical neighbors, suggesting horizontal transmission.

If a trait is mainly transmitted vertically, this implies that population movements probably accompanied the spread of this trait. It has been extensively debated whether the spread of agriculture across Europe at the beginning of the Neolithic was accompanied by a population expansion of farmers (Ammerman and Cavalli-Sforza, 1984; c.f. Zvelebil, 1995). We can ask the same question about the spread of domesticated livestock, for example, across Africa.

### Methods and results

There were two parts to the analysis. First, we tested whether the variables in the analysis were transmitted vertically and/or horizontally between populations. Second, we tested the three major hypotheses for variation in the prevalence of lactase persistence across populations. A world-wide cross-cultural sample of 62 populations was used. Past relationships among these populations were estimated using a genetic distance tree from Cavalli-Sforza et al. (1994).

### Data

The populations in the sample were taken from comprehensive reviews by Simoons (1978), Flatz (1987) and Sahi (1994) and from a literature search of the BIDS science citation index from 1981–96 using the search terms "lactose absorption" and "lactose malabsorption." These sources were intended to include all populations sampled for lactase persistence.

Additional criteria for inclusion in the sample were as follows. Only samples from adults were included, unless otherwise stated in Table 12.1. All populations in the sample were also included in Cavalli-Sforza *et al.* (1994), enabling them to be placed on a genetic tree. Cavalli-Sforza *et al.* (1994) only included "aboriginal" populations, defined as populations that inhabited approximately their present location in 1492. Thus, non-Native Americans and non-Aboriginal Australians (among others) were excluded from the dataset. Populations recognized to have mixed ancestry were also excluded. Other migrant populations sampled for lactase persistence were also omitted, to decrease the probability of recent genetic admixture in the dataset.

All populations in the sample are also found in the *Ethnographic Atlas*, a cross-cultural database compiled by Murdock (1967), currently in the process of being revised and computerised by P. Gray (Murdock, no date). Any ambiguities in matching populations sampled for lactase persistence to cultures in the *Ethnographic Atlas* are noted in Table 12.1. Data on pastoralism and the geographical location of populations in the sample were taken from the *Ethnographic Atlas*. Australian Aborigines were an exception, not being a single culture, as recognized by the *Ethnographic Atlas*. However, they were a genetically monophyletic group, so could be placed on the genetic tree (Cavalli-Sforza *et al.*, 1994, p. 78). They were traditionally hunter-gathers without domesticated livestock.

In total 7905 individuals from 62 distinct cultures were included in the dataset, shown in Table 12.1. The greatest loss of individual samples resulting from the selection criteria outlined above was from the non-aboriginal populations of America and Australia. More importantly for the aims of this study, a number of samples from "anthropological" populations were unable to be used because these populations are not known genetically, or are not included in the *Ethnographic Atlas*. Anthropologically interesting populations lost included, among others, the hunter-gatherer Khants from Western Siberia (94% lactose maldigesters, Lember *et al.*, 1995) and several other groups from the former Soviet Union (Sahi, 1994); the pastoralist Kasakhs (Wang *et al.*, 1984); the Roma (Gypsies), (56% lactose maldigesters; Czeizel *et al.*, 1983); and various

Indian groups (45% lactose maldigesters in North India, 67% in South India; Tandon *et al.*, 1981). The inclusion criteria applied here, however, had the advantage of allowing all populations to be compared using the same source for variables such as pastoralism, which had been previously quantified by an independent researcher. It also allowed the populations in the sample to be placed on a genetic population tree, so that independent instances of evolutionary change in variables could be identified.

The *Ethnographic Atlas* includes information on the importance of livestock in each population's total subsistence activities, given as the percent of total subsistence that comes from pastoralism. The *Ethnographic Atlas* codes pastoralism as a quantitative trait (0–100% dependence on pastoralism) divided into 10 classes. The midpoint of the *Ethnographic Atlas* estimate of dependence on pastoralism was used here (i.e., if the *Ethnographic Atlas* scored a culture's dependence on livestock as 3, or 26–35% of total subsistence activities, this was estimated to be 30.5% dependence on pastoralism here). The *Ethnographic Atlas* also includes information about the main type of livestock kept, and whether or not milking was traditionally practiced. "Pastoralism" is used here only in reference to livestock capable of being milked. If the main type of animals kept was pigs or small domestic animals (e.g., dogs), the culture was recorded as "non-pastoralist" here (zero dependence). The analysis was repeated twice, first including, then excluding, cultures that traditionally did not milk their animals.

No distinction was made between populations that consume predominantly processed, low-lactase forms of milk and populations that consume significant amounts of fresh milk, with a high lactose content. Ethnographic evidence suggests that milk-processing technologies are present in all dairying or pastoralist groups today. In hot climates in Africa and the Middle East, milk is soured naturally if it is left to stand. It seems probable that milk-processing technologies were present very early in the history of milking domestic animals. It was therefore assumed here that all pastoralists have had equally effective milk-processing technology, whether they lived 6000 years ago or more recently. This assumption contrasts with Durham (1991), who interpreted the high frequencies of high lactose digestive capacity in some present-day North African pastoralists as the outcome of their having adopted a pastoralist mode of subsistence very early, before the full development of milk-processing technologies.

Aridity was estimated by the number of months per year with less than 50 ml rainfall in the area inhabited by each population (Pearce and Smith, 1993). Other measures of aridity were tried, including average annual rainfall, and average rainfall in environments above 30 °C, but this made no difference to the outcome of the analysis. The amount of sunshine experienced by each population was estimated from the global solar radiation for land at that longitude and latitude (Essenwanger, 1985). Latitude and longitude were taken from the *Ethnographic Atlas*.

The genetic tree used to estimate the past relationships among populations was adapted from Cavalli-Sforza *et al.* (1994). It was based on their $F_{ST}$ linkage trees (shown in Figures 12.1 and 12.2). It was assumed that all branch lengths were equal.

### *The geographical and phylogenetic distribution of lactase persistence and pastoralism*

In Figure 12.1, lactase persistence has been mapped onto the genetic tree using parsimony. Figure 12.3 shows lactase persistence in populations in the sample on a map. High levels of lactase persistence ( > 70%) are found in Northern Europe, represented here by the Irish, and in some other populations in West and Central Europe (the Spanish and Czechs). High levels of lactase persistence are also found in some pastoralist populations in Africa and the Middle East, including the Bedouin (camel herders in the Middle East and North Africa), nomadic Fulani and Tuareg (pastoralists in the Sahara-Sahel) and the traditionally cattle-herding Tutsi in East Africa. Intermediate levels of lactase persistence (30–70%) are found in populations elsewhere in Europe and in parts of South Asia, in Afghanistan, Pakistan and India. In Northern Europe, the Sámi (also known as Lapps) have intermediate levels of lactase persistence. Intermediate levels of lactase persistence are also found in some African pastoralist or agro-pastoralist populations, including the Baggara nomads and Nubian pastoralists in Northeast Africa, and the Sotho and Nama in Southern Africa. Low frequencies of lactase persistence ( < 30%) are found throughout the rest of the world, including many African populations in East and Southeast Asia, the Pacific, and in Australian Aborigines and Native Americans. Further descriptions of world-wide patterns in lactase persistence can be found in Simoons (1978), Flatz (1987) and Sahi (1994).

On the genetic tree (Figure 12.1) it can be seen that high levels of lactase persistence are mostly clustered within a clade that includes Europeans, South and West Asians and North Africans. Independent instances of the evolution of high levels of lactase persistence occur scattered throughout the sub-Saharan African clade. As expected from the geographical distribution of this trait, lactase persistence is virtually absent in the Southeast Asian, East Asian and Native American clades.

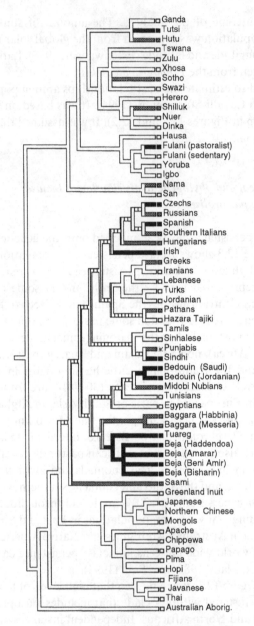

Figure 12.1. Genetic tree used as a model of past relationships among
populations, adapted from Cavalli-Sforza *et al.*, (1994). Lactase persistence has
been mapped onto the tree using parsimony (Maddison and Maddison, 1992).
Frequencies of lactase persistence (LP) are grouped into three groups for
illustrative purposes only: black, >70% LP; grey, 30–70% LP; white, <30% LP.
Hatched lines show ambiguous ancestral states.

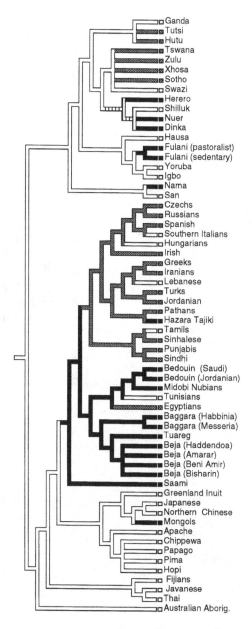

Figure 12.2. Genetic tree used as a model of past relationships among populations, adapted from Cavalli-Sforza *et al.*, (1994). Dependence on pastoralism has been mapped onto the tree using parsimony (Maddison and Maddison, 1992). Pastoralism has been grouped into three groups for illustrative purposes only: black, > 50% dependence; grey, 30–50% dependence; white, < 30% dependence.

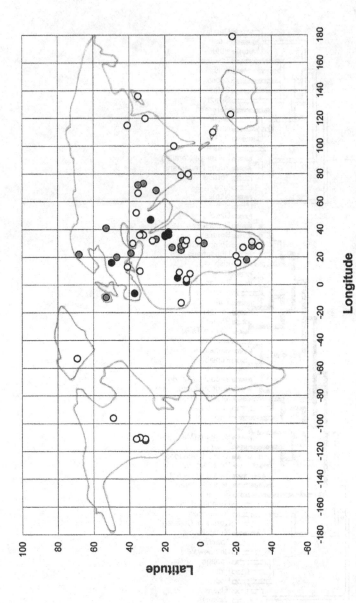

Longitude

● >70% Lactase persistent
◉ 30%–70% Lactase persisent
○ <30% Lactase persistent

Figure 12.3.  Lactase persistence in 62 populations worldwide.

Dependence on livestock is shown on a map in Figure 12.4, and on the genetic tree in Figure 12.2. A comparison of the two genetic trees, and the two maps, shows that the pattern of livestock dependence is broadly similar to the distribution of lactase persistence, but that the former is more widespread than the latter. The root of the clade that includes Europeans, South and West Asians and North Africans shows high dependence on livestock. This is consistent with archaeological evidence of the antiquity of domestic livestock in the Middle East and North Africa (Sherratt, 1980; Clutton-Brock, 1987). In sub-Saharan Africa, the root of the Bantu- and Nilotic-speaking clade shows moderate to high dependence on livestock, but pastoralism is absent from higher nodes (towards the root) within the sub-Saharan African part of the tree. Again, this is consistent with archaeological evidence.

### The transmission of traits between populations

The first part of the statistical analysis was to investigate whether the variables in each population are more similar to that population's geographical or phylogenetic neighbors. This is important in determining how traits are likely to be transmitted, and also whether phylogenetically controlled statistics are needed in tests of coevolutionary hypotheses. Using multiple regression, we tested whether each trait in the sample was associated with the populations' phylogenetic relatives, and/or geographical neighbors. The dependent variable in the multiple regression was the trait whose mode of transmission was being tested, in each population in the sample. The two independent variables were (a) that trait in the populations' phylogenetic relatives, and (b) that trait in the populations' geographical neighbors. If a trait is associated with phylogenetic relatives, this indicates that it is transmitted vertically, from "mother" to "daughter" populations. If a trait is associated with nearest geographical neighbours this suggests that it is transmitted horizontally, between neighboring populations.

Phylogenetic relatives were defined as the sister-group of each population on the genetic tree (Figures 12.1, 12.2). Most populations could be compared to a single other population, with which they formed a paired clade at the tips of the tree (e.g., the Czechs and Russians). In cases where three or more populations descended from a single node (e.g., the Shilluk, Dinka and Nuer), each population was compared with the mean value of all the other populations in the clade. In some cases (e.g., the Sámi and Australian Aborigines) a single population was coordinate with a larger

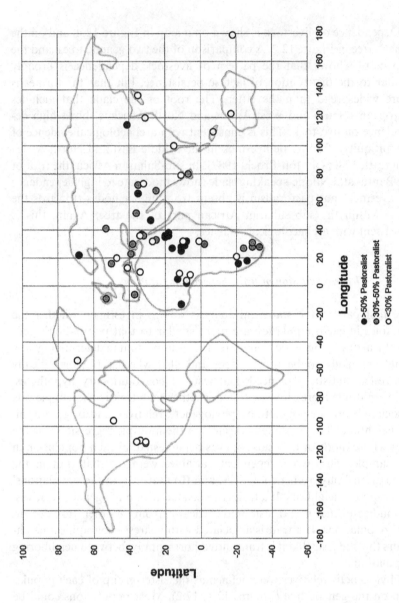

Figure 12.4. Dependence on pastoralism in 62 populations worldwide.

Table 12.2. *Similarity between populations in biocultural traits, according to geographical proximity or ancestry*

| Dependent variable | Phylogenetic relative(s) | | | Geographical neighbor(s) | | |
|---|---|---|---|---|---|---|
| | Slope | S.E. of slope | Significance of slope | Slope | S.E. of slope | Significance of slope |
| Lactase persistence | 0.497 | 0.163 | 0.004 | 0.102 | 0.165 | 0.6 (n.s) |
| Pastoralism | 0.520 | 0.152 | 0.002 | 0.142 | 0.149 | 0.4 (n.s) |
| Aridity | 0.396 | 0.149 | 0.01 | 0.385 | 0.157 | 0.02 |
| Solar radiation | 0.515 | 0.185 | 0.01 | 0.501 | 0.211 | 0.03 |

*Note:* Each line represents a separate multiple regression, in which the dependent variable was the trait in each population in the sample, and the independent variables were that trait in each population's closest phylogenetic relative and nearest geographical neighbor. The significance values shown are of partial regression coefficients. All multiple regressions were highly significant overall ($F$ significant at $p < 0.0002$).

clade. In these cases, the population was compared to the mean of all populations in the coordinate clade.

Nearest neighbors were found using great-circle distances. Each population was compared with the same number of phylogenetic relatives and geographical neighbors. Thus, populations which were compared with a single geographical relative were also compared with a single geographical neighbor (the nearest population). Populations compared with two or more phylogenetic relatives were compared with an equal number of their nearest geographical neighbors. Where populations were compared with more than one geographical neighbor, the mean value of each trait in all the geographical neighbors was used.

Results of the test of the transmission of traits among populations are shown in Table 12.2. Lactase persistence and pastoralism are highly significantly associated with population history (phylogeny), but not with proximity to neighboring populations. This suggests that these traits are mostly transmitted vertically, from "mother" to "daughter" populations. Aridity and solar intensity are associated with both phylogeny and geographical neighbors, suggesting that these environmental variables are similar among neighboring populations, as one would expect, and also that daughter populations probably tend to inhabit similar environments to their ancestors. Because these environmental variables are related to latitude, and nearest geographical neighbors are defined using longitude and latitude, the association with geographical neighbors is expected.

### Phylogenetic comparative analysis

Two phylogenetic comparative methods were used to test the relationship between lactase persistence, pastoralism, aridity and solar intensity. The first was Felsenstein's (1985) method of comparative analysis using independent contrasts. The second was Pagel's (1994) maximum likelihood method for testing for correlated evolution in binary discrete characters.

#### Comparative analysis using independent contrasts

Felsenstein's (1985) method of comparative analysis using independent contrasts tests for correlated evolution among continuous variables, on a phylogenetic tree. Populations in the sample were placed on a genetic tree, used as a model of the ancestral relationships among populations. Variables were mapped onto the tree. If equal branch lengths are assumed, the character state at the higher node is equivalent to the mean of the character states at the immediately lower nodes. Evolutionary change in variables along the tree's branches was measured using "independent contrasts," which are differences in character states among daughter populations descending from a single node, or between nodes descending from a higher node (Felsenstein, 1985). A model of evolution by Brownian or random motion was used, in which the net change in a character along a branch is expected to be zero, with variance proportional to the branch length (Felsenstein, 1985; Pagel, 1992; Purvis and Rambaut, 1995). Correlated evolutionary change among variables along the branches of the tree was tested for. For a bifurcating tree, the number of independent contrasts is equal to the number of populations in the sample minus one. Unresolved nodes were expanded using the method of Pagel (1992), using pastoralism (the main independent variable) as the special variable. Each independent contrast contributes one degree of freedom when calculating the significance of relationships between variables. This method was implemented using the computer program CAIC (Purvis and Rambaut, 1995).

The relationship between independent contrasts in lactase persistence, pastoralism, aridity and solar intensity was tested using multiple regression through the origin, using SPSS (Norušis, 1994). The regression is through the origin, because independent contrasts are a measure of differences among sister populations, not actual character values (Felsenstein, 1985; Pagel, 1992). The dependent variable was independent contrasts in lactase persistence.

Results of the comparative analysis using independent contrasts are shown in Table 12.3. Lactase persistence was significantly associated with

Table 12.3. *Multiple regression, through the origin, of independent contrasts*

| (a) Overall significance of multiple regression model | |
|---|---|
| Multiple $R$ | 0.51 |
| $R^2$ | 0.26 |
| Significance ($R$) | 0.003 |

| (b) Individual variables | | | |
|---|---|---|---|
| | Slope | S.E. slope | Significance |
| Pastoralism | −0.489 | 0.148 | 0.002 |
| Solar radiation | −0.255 | 0.186 | 0.2 (n.s.) |
| Aridity | −0.446 | 1.117 | 0.7 (n.s.) |

*Note:* The frequency of lactase *non*-persistence in each population was used (hence the negative relationship with dependence on pastoralism). Significance values of the slopes are *t*-values.

pastoralism, but not with aridity or solar intensity. This suggests that lactase persistence and pastoralism evolve together, but that aridity and solar intensity are not additional selective pressures for the evolution of lactase persistence.

It did not make a significant difference to the results whether the populations which traditionally kept livestock but did not milk their animals were counted as pastoralists or not. Counting these populations as non-pastoralist slightly increased the association between pastoralism and lactase persistence, as the milk-drinking hypothesis predicts (Simoons, 1979) but the significance level of the result was not changed. These populations formed a very small proportion of the populations in the dataset.

*Maximum likelihood test for correlated evolution in binary characters*

Pagel's (1994) method for testing for correlated evolution among discrete binary characters was used to test whether high levels of lactase persistence are more likely to evolve in populations that milk their livestock. This method uses maximum likelihood. Unlike comparative analysis using independent contrasts, it does not rely on a single set of reconstructed character states. Instead, it represents all character states at internal nodes on the tree as probability distributions. Therefore, all possible transitions in character states on the phylogeny can be taken into account when calculating the likelihood of different models of evolution. This method also allows hypotheses about the direction of evolutionary change to be tested, in addition to hypotheses about correlated evolution.

This method tests the hypothesis that evolutionary change in two variables is correlated by comparing the fit of two models of evolution to the data on the tree. Evolution is modeled by a Markov process, in which the probability of change in a character is dependent on the present state of the character. The first model is an independent model, in which the two characters evolve independently along the branches of the tree. The second model is a dependent model, in which the probability of change in one character is dependent on the state of the other character. The goodness of fit of the two models is compared using a likelihood ratio (LR) statistic, defined as $LR = -2\log_e[L(I)/L(D)]$, where $L(I)$ is the likelihood of the model of independent evolution and $L(D)$ is the likelihood of the model of dependent evolution. If the model of correlated evolution is significantly more likely, this is evidence that evolution in the two characters is correlated (Pagel, 1994; 1998).

Pagel's (1994) method tests for correlated evolution among discrete binary variables, so lactase persistence was divided into two classes, high and low prevalence, following its bimodal distribution among populations in the sample (Figure 12.5). High prevalence of lactase persistence is defined as more than 70% of individuals being lactase persistent. Milking is also a two-state discrete character (presence or absence of milking). Data on the presence of milking were taken from the *Ethnographic Atlas*. We tested whether lactase persistence was more likely to change from low to high if milking was present.

This method can also be used to estimate the probable direction of evolutionary change in two characters. The importance of each transition in the model of correlated evolution is assessed by comparing the goodness of fit of the unrestricted model of correlated evolution (the full dependent model), with a restricted model, in which the rate of one of the eight possible transitions has been set to zero. The restricted and unrestricted models are compared using a likelihood ratio statistic. If the goodness of fit of the model of restricted evolution is significantly less good than the full model, this is evidence that this transition is significant. We tested the significance of each evolutionary transition in the dependent model of correlated evolution (Figure 12.6).

Pagel's (1994) method was implemented using the computer program DISCRETE (available from M. Pagel, School of Animal and Microbial Sciences, University of Reading, UK). This program requires a fully

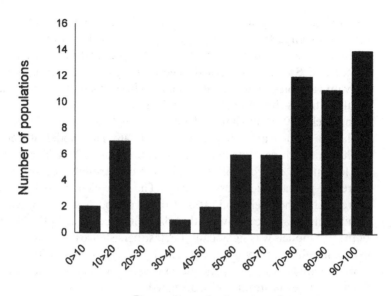

Figure 12.5. Bimodal distribution of lactase persistence across populations in the sample.

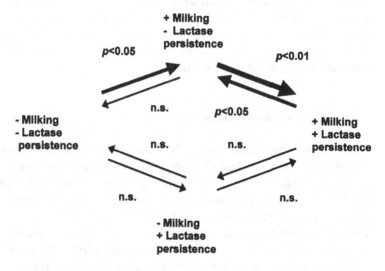

Figure 12.6. The direction of change in the coevolution of milking and lactase persistence. The significance of transitions between character states are shown, calculated using Pagel's (1994) maximum likelihood method. The significance of each evolutionary transition in the model is shown next to the arrows.

resolved bifurcating phylogeny. Unresolved nodes on the genetic tree were resolved by linguistic relationships where possible (following the linguistic classifications of Ruhlen, 1991). The Beja populations, and the Zulu, Xhosa and Swazi were pooled. These groups had no variation in the relevant variables. It was assumed that all branch lengths were equal. The resulting tree is shown in Figures 12.7 and 12.8.

In Pagel's (1994) maximum likelihood model, the model of independent evolution yielded a log-likelihood of −41.46. The model of dependent evolution for these traits had a log-likelihood of −37.04. These numbers are logarithms of probabilities, and the more strongly negative numbers represent probabilities closer to zero. The dependent model is therefore more likely. The likelihood ratio of the two models was 8.83, which was significant at $p < 0.05$, showing that the difference between the likelihoods of the two models was significant. Lactase persistence is thus more likely to evolve in milking populations. This finding corresponds with the finding from the analysis by independent contrasts that the evolution of lactase persistence and pastoralism is correlated.

The significance of each evolutionary transition is shown on Figure 12.6. From an ancestral condition of no milking (−milking) and low levels of lactase persistence (−LP), it is probable that milking (+milking) evolved first, followed by high levels of lactase persistence (+LP). In some cases, populations with milking that evolved high levels of lactase persistence may have subsequently lost high levels of lactase persistence. It appears that high levels of lactase persistence never evolved without the prior presence of milking. This further supports the hypothesis that lactase persistence is an adaptation to dairying.

## Discussion and conclusions

The analysis of the transmission of traits among populations showed that the four traits hypothesized to be related (lactase persistence, pastoralism, aridity and solar intensity) cluster in a non-random way with respect to phylogeny. It is therefore necessary to control for phylogeny when testing hypotheses about the relationships between these variables across populations. Interestingly, pastoralism, a cultural trait, showed a strong association with phylogeny, but no additional association with geographical neighbors. It is often assumed that cultural or behavioral traits are likely to be transmitted by horizontal diffusion, but this does not seem to be the case for pastoralism. The phylogenetically conserved character of pastoralism was also found using a different method by Guglielmino *et al.* (1995). Lactase persistence also appears to be predicted by a population's

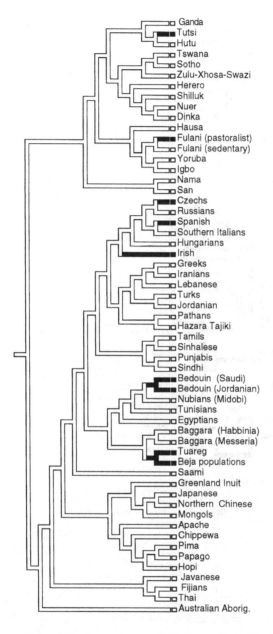

Figure 12.7. Composite tree used in the maximum likelihood analysis. Tree is based on Cavalli-Sforza *et al.*, (1994) $F_{ST}$ distance tree. Unresolved nodes were resolved using linguistic relatedness. Lactase persistence (LP) is mapped onto the tree using parsimony: the most parsimonious estimate of ancestral states is only one possible reconstruction considered in this method. Black indicates LP > 70%, white indicates LP < 70%.

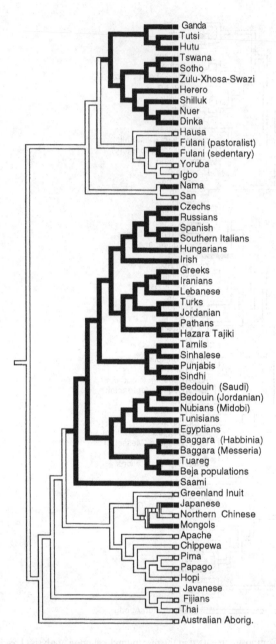

Figure 12.8. Composite tree used in the maximum likelihood analysis. Milking has been mapped onto the tree using parsimony: the most parsimonious estimate of ancestral states is only one possible reconstruction considered in this method. Black indicates milking populations, white indicates non-milking populations.

ancestry, but not by the frequency of lactase persistence in its geographical neighbors.

The results of the phylogenetic comparative analysis support the hypothesis that lactase persistence evolves in populations which rely on domestic livestock for their subsistence. The results do not support Flatz and Rotthauwe's (1973) hypothesis that fresh milk consumption and lactase persistence are additionally selected for at high latitudes with limited sunlight, or Cook and Al-Torki's (1975) hypothesis that lactase persistence is additionally selected for in hot, arid environments. Aridity and pastoralism are related, because pastoralism is often adopted as a means of subsistence in regions where rainfall is too low for agriculture, but no evidence was found here that aridity *per se* provides an additional selective pressure for the evolution of lactase persistence. The analysis using Pagel's (1994) maximum likelihood model shows that the evolution of lactase persistence is associated with milking, and further shows that milking is adopted before lactase persistence evolves.

An historical perspective on the spread of domesticated livestock explains why some pastoralists, for example the Herero in Southern Africa, have a low prevalence of lactase persistence. These populations may have become pastoralists too recently for the gene for lactase persistence to have spread. The archaeological evidence suggests that pastoralism in Southern Africa dates from the 1st millennium AD (Sherratt, 1980; Bower, 1995). This is much more recent than pastoralism in North Africa and the Middle East, where pastoralists with a higher prevalence of lactase persistence live (for example, the Bedouin and the Tuareg). Other questions remain unanswered, for example, why lactase persistence is rare in some Sudan-Sahel populations (e.g., the Nuer and Dinka) in a region with an ancient pastoralist tradition. Gene flow from non-pastoralist populations may have occurred, or these populations may be relatively recent immigrants to the region.

The evolution of adult lactase persistence has been modeled several times. Bodmer and Cavalli-Sforza (1976) estimated that a selection coefficient of 0.04 would be necessary for lactase persistence to increase from an initial prevalence of 0.001% to the levels observed today in Northern European populations (estimated frequency 0.5) within 290 generations (9000 years). If the initial frequency of the lactase persistence gene was 1.0%, a selection coefficient of only 0.015 would be required. This time scale is realistic for the Middle East where livestock were first domesticated at around 8000–7000 BCE (Clutton-Brock, 1987). Flatz (1987) estimated that for the gene to reach contemporary European levels in the 3500 years or less since the first known domestic livestock in northern Europe, starting from an initial frequency of 0.005%, a higher selection coefficient of between 3% and 7% would be required. This calculation was based on the assumption that pastoralism spread by diffusion to indigenous hunter-gatherers in Europe, rather than being introduced by colonization by Middle Eastern pastoralists (c.f. Ammerman and Cavalli-Sforza, 1984).

More recently, attempts have been made to model the coevolution of a gene for lactase persistence and the cultural trait of milk drinking. Aoki (1986) estimated that for the selection of the gene for lactase persistence from an initial prevalence of 0.05% to the prevalence observed in Northern Europe today (estimated gene frequency 0.7), within the time available since the advent of dairying (6000 years), and with an effective population size of 500, the selection coefficient must have been greater than 5%. Feldman and Cavalli-Sforza (1989) also found that a selection coefficient of greater than 5% was necessary for a gene frequency of 0.7 to be reached in 6000 years.

However, in these dual-inheritance, coevolutionary models, milk drinking is a cultural trait with a low initial frequency, whose selection coefficient is dependent on the prevalence of the lactase persistence gene. The ethnographic evidence does not support this assumption, insofar as milk consumption apparently has been universally adopted within some predominantly lactase non-persistent populations, for example the Mongols, the Herero, the Nuer and the Dinka (Table 12.1). Milk-based pastoralism may be the best means of subsistence in dry, marginal environments, even for lactose non-digesters. Milk processing, and the consumption of fresh milk only in small quantities, are cultural and behavioral means by which many lactase non-persistent individuals manage to consume milk products without suffering the symptoms of lactose intolerance. After milk-based pastoralism has been adopted as a means of subsistence, lactase persistence would be selected for, because it would enable adults to consume more fresh milk and to derive a nutritional benefit from the lactose component of fresh milk. The "initial" frequency of the cultural trait of milk consumption may therefore be virtually 100%, which could reduce the selection coefficient required for the lactase persistence gene to reach observed frequencies in the time available. In this case these traits would not be truly coevolutionary, because selection for milk consumption would not be dependent on the gene for lactase persistence.

It may appear surprising that low solar intensity (at high latitudes) was not an additional selective pressure for the evolution of lactase persistence, when lactase persistence is found at high levels in Northern Europe, yet domestic animals only represent about 30% of total subsistence in these regions today (Murdock, 1967). Pastoralists with similar levels of lactase persistence in Africa and the Middle East have much higher levels of dependence on animals, and the origin of domesticated animals in North Africa and the Middle East is also thought to be more ancient (Sherratt, 1980). Archaeological evidence for subsistence practices in Northern Europe in the Neolithic may shed some light on the Northern European puzzle. Liden (1995) carried out a dietary study of two Swedish Neolithic populations, testing nitrogen and carbon isotopes in skeletons. The nitrogen isotopes indicated a high level of meat consumption, indicating that the Neolithic populations were pastoralists, rather than mixed farmers like more recent Northern Europeans. If the ancestors of modern Northern Europeans were pastoralists, this could explain the high prevalence of lactase persistence in Northern Europe today. This is assuming that a high meat diet also indicates that milk was consumed, which may not have been the case: Sherratt (1981) argued that the "secondary products revolution" involving milk and wool production post-dated the origin of domesticated animals by several thousand years.

It would not be surprising if the level of reliance on domesticated livestock has changed since the Neolithic in Europe. What is perhaps surprising is that reliance on domesticated livestock has been sufficiently constant within populations, over millennia, for contemporary ethnographic data to have explanatory power over an evolutionary process that began thousands of years ago, selection for lactase persistence in pastoralist populations. As shown above, pastoralism is highly significantly associated with population history, suggesting that this cultural trait may be a highly conserved down the generations. Pastoralism appears to have been stable enough for the evolution of lactase persistence to occur, and for this evolutionary relationship still to be detectable in extant populations.

Some aspects of "tree" models of population history will now be discussed. A tree is a simplified model of the past relationships among populations. Gene frequency trees reflect actual population history accurately if population size has been constant, and there has been no selection, gene flow or admixture. Clearly, these assumptions were violated to a varying extent throughout human population history. Admixture, referring to merging between previously divergent populations, requires a network model, with joining between branches (anastomoses) rather than a tree. On a tree, admixed populations are joined as outliers to the parental branch that contributed most to their ancestry (Cavalli-Sforza *et al.*, 1994). As stated above, recently admixed populations were excluded from the dataset in this analysis. The effect of gene flow on genetic population trees is to shorten the branch lengths. If gene flow is homogenous across the tree, branch lengths will retain the same proportions within the tree. If the amount of gene flow is variable across the tree, parts of the tree with high gene flow will appear younger relative to parts of the tree with low gene flow (Weiss, 1996). The effects of demographic processes on genetic population trees are discussed further in Holden (1999).

Assumptions about the relationships between populations are always implicit in a comparative analysis, even if they are not made explicit. In a standard cross-population comparison, an implicit assumption is that the populations in the sample are equidistantly related, i.e., in a "star" phylogeny (Felsenstein, 1985; Pagel and Harvey, 1989). Although one is often uncertain which tree is the best model of past relationships among human populations (or whether a network might be preferable to a tree) it is certain that genetic trees provide a better estimate of the hierarchical relationships among populations than a star phylogeny. Unresolved nodes and inconsistencies between different trees highlight our lack of knowledge about the past relationships among populations, showing where new data should be gathered. A task for the future is the development of models of population history that incorporate anastomoses (admixture) and porous boundaries between populations. We have shown elsewhere (Holden and Mace, 1997) that the results presented here were also obtained using alternative trees, including another genetic tree and a language tree. This increases our confidence that this result is not an artifact of a particular tree.

### Acknowledgments

We thank the Royal Society, the Leverhulme Trust and NERC for funding this research. We thank Mark Pagel for making the program DISCRETE available to us. We are grateful to Rob Baulk for his method of calculating great-circle distances, and to Sophie Holden for her help with the analysis.

### References

Ammerman, A.J. and Cavalli-Sforza, L.L. (1984). *The Neolithic transition and the genetics of population in Europe.* Princeton, NJ: Princeton University Press.
Aoki, K. (1986). A stochastic model of gene-culture coevolution suggested by the "culture historical hypothesis" for the evolution of lactose absorption in humans. *Proceedings of the National Academy of Sciences USA* **83**, 2929–2933.
Bodmer, W.F. and Cavalli-Sforza, L.L. (1976). *Genetics, Evolution and Man.* San Francisco: W.H. Freeman.
Bower, J. (1995). Early food production in Africa. *Evolutionary Anthropology* **4**, 130–139.

Cavalli-Sforza, L.L., Menozzi, P. and Piazza, A. (1994). *The History and Geography of Human Genes*. Princeton, NJ: Princeton University Press.

Clutton-Brock, J. (1987). *A Natural History of Domesticated Mammals*. London: British Museum (Natural History) and Cambridge: Cambridge University Press.

Cook, G.C. (1978). Did persistence of intestinal lactase into adult life originate on the Arabian peninsula? *Man* (New Series) **13**, 418–427.

Cook, G.C. and Al-Torki, M.T. (1975). High intestinal lactase concentrations in adult Arabs in Saudi Arabia. *British Medical Journal* **III**, 135–136.

Czeizel, A., Flatz, G. and Flatz, S.D. (1983). Prevalence of primary adult lactose malabsorption in Hungary. *Human Genetics* **64**, 398–401.

Durham, W. (1991). *Coevolution: Genes, Culture and Human Diversity*. Stanford, CA: Stanford University Press.

Essenwanger, O.M., ed. (1985). *World Survey of Climatology: General Climatology 1A*. Amsterdam: Elsevier.

Feldman, M.W. and Cavalli-Sforza, L.L. (1989). On the theory of evolution under genetic and cultural transmission with application to the lactose absorption problem. In *Mathematical Evolutionary Theory*. ed. M.W. Feldman, pp. 145–173. Princeton, NJ: Princeton University Press.

Felsenstein, J. (1985). Phylogenies and the comparative method. *American Naturalist* **125**, 1–15.

Flatz, G. (1987). Genetics of lactose digestion in humans. *Advances in Human Genetics* **16**, 1–77.

Flatz, G. and Rotthauwe, H.W. (1973). Lactose nutrition and natural selection. *Lancet* **II**, 76–77.

Guglielmino, C.R., Viganotti, C., Hewlett, B. and Cavalli-Sforza, L.L. (1995). Cultural adaptation in Africa: role of mechanisms of transmission and adaptation. *Proceedings of the National Academy of Sciences USA* **92**, 7585–7589.

Harvey, P.H. and Pagel, M.D. (1991). *The Comparative Method in Evolutionary Biology*. New York: Oxford University Press.

Holden, C. (1999). *The evolution of human diversity: a phylogenetic approach*. Unpublished PhD thesis, University College London.

Holden, C. and Mace, R. (1997). Phylogenetic analysis of the evolution of lactose digestion in adults. *Human Biology* **69**, 605–628.

Holden, C. and Mace, R. (1999). Sexual dimorphism in stature and women's work: a phylogenetic cross-cultural analysis. *American Journal of Physical Anthropology* **110**, 27–45.

Johnson, J.D., Simoons, F.J., Hurwitz, R., Grange, A., Mitchell, C.H., Sinatra, F.R., Sunshine, P., Robertson, W.V., Bennett, P.H. and Kretchmer, N. (1977). Lactose malabsorption among Pima Indians of Arizona. *Gastroenterology* **73**, 1299–1304.

Lember, M., Tamm, A. and Villako, K. (1991). Lactose malabsorption in Estonians and Russians. *European Journal of Gastroenterology and Haematology* **3**, 479–481.

Lember, M., Tamm, A., Piirsoo, K., Suurmaa, K., Kermes, K., Kermes, R., Sahi,

T. and Isokoski, M. (1995). Lactose malabsorption in Khants in Western Siberia. *Scandanavian Journal of Gastoenterology* **30**, 225–227.

Liden, K. (1995). Megaliths, agriculture, and social complexity: a diet study of two Swedish megalith populations. *Journal of Anthropological Archaeology* **14**, 404–417.

Mace, R. and Holden, C. (1999). Evolutionary ecology and cross-cultural comparison: the case of matrilineal descent in sub-Saharan Africa. In *Comparative Primate Socioecology*. ed. P.C. Lee, pp. 387–405. Cambridge: Cambridge University Press.

Mace, R. and Pagel, M. (1994). The comparative method in anthropology. *Current Anthropology* **35**, 549–564.

Maddison, W.P. and Maddison, D.R. (1992). *MacClade: Analysis of Phylogeny and Character Evolution. Version 3.0.* Sunderland, MA: Sinauer Associates.

Mayr, E. (1963). *Animal Species and Evolution.* Cambridge, MA: Belknap Press.

McCracken, R.D. (1971). Lactase deficiency: An example of dietary evolution. *Current Anthropology* **12**, 479–500.

Metneki, J., Czeizel, A., Flatz, S.D. and Flatz, G. (1984). A study of lactose absorption capacity in twins. *Human Genetics* **67**, 296–300.

Murdock, G.P. (1967). *Ethnographic Atlas.* Pittsburgh: University of Pittsburgh Press.

Murdock, G.P. (No date). Revised computerized version of Murdock's *Ethnographic Atlas.* Personal communication from P. Gray via D. Sellen of the Anthropology Department, University College London.

Mustapha, A., Hertzeler, S.R. and Savaiano, D.A. (1997). Lactose: nutritional significance. In *Advanced Dairy Chemistry*, Vol. 3. *Lactose, Water, Salts and Vitamins*, ed. P.F. Fox, pp. 127–154, 2nd edn. London: Chapman and Hall.

Norušis, M.J. (1994). *SPSS for Windows.* Version 6.1. Chicago: SPSS.

Pagel, M. (1992). A method for the analysis of comparative data. *Journal of Theoretical Biology* **156**, 431–442.

Pagel, M. (1994). Detecting correlated evolution on phylogenies: a general method for the comparative analysis of discrete characters. *Proceedings of the Royal Society of London (B)* **255**, 37–45.

Pagel, M. (1998). Inferring evolutionary processes from phylogenies. *Zoologica Scripta* **26**, 331–348.

Pagel, M. and Harvey, P. (1989). Comparative methods for examining adaptation depend on evolutionary models. *Folia Primatologia* **53**, 203–220.

Pearce, E.A. and Smith, C.G. (1993). *The World Weather Guide.* Oxford: Helicon Publishing.

Purvis, A. and Rambaut, A. (1995). Comparative analysis by independent contrasts (CAIC): an Apple Macintosh application for analysing comparative data. *Computer Applications for the Biosciences* **11**, 247–251.

Ransome-Kuti, O., Kretchmer, N., Johnson, J.D. and Gribble, J.T. (1975). A genetic study of lactose digestion in Nigerian families. *Gastroenterology* **68**, 431–436.

Ruhlen, M. (1991). *A Guide to the World's Languages.* Vol. 1. *Classification*, 2nd edn. London: Edward Arnold.

Sahi, T. (1994). Genetics and epidemiology of adult-type hypolactasia. *Scandinavian Journal of Gastroenterology* (Suppl 29) **202**, 1–6.
Sahi, T., Isokoski, J., Jussila, K., Launiala and Pyorala, K. (1973). Recessive inheritance of adult-type lactose malabsorption. *Lancet* **II**, 823–826.
Scrimshaw, N.S. and Murray, E.B. (1988). The acceptability of milk and milk products in populations with a high prevalence of lactose intolerance. *American Journal of Clinical Nutrition* (Suppl) **48**, 1083–1159.
Sherratt, A., ed. (1980). *Cambridge Encyclopaedia of Archaeology*. Cambridge: Cambridge University Press.
Sherratt, A. (1981). Plough and pastoralism: aspects of the secondary products revolution. In *Pattern of the Past*. ed. I. Hodder, G. Isaac and N. Hammond, pp. 261–306. Cambridge: Cambridge University Press.
Simoons, F.J. (1969). Primary adult lactose intolerance and the milking habit: a problem in biological and cultural interrelations. I. Review of the medical research. *American Journal of Digestive Diseases* **14**, 819–836.
Simoons, F.J. (1970a). Primary adult lactose intolerance and the milking habit: a problem in biological and cultural interrelations. II. A culture historical hypothesis. *American Journal of Digestive Diseases* **15**, 695–710.
Simoons, F.J. (1970b). The traditional limits of milking and milk use in Southern Asia. *Anthropos* **65**, 547–593.
Simoons, F.J. (1978). The geographic hypothesis and lactose malabsorption: a weighing of the evidence. *American Journal of Digestive Diseases* **23**, 963–980.
Simoons, F.J. (1979). Dairying, milk use, and lactose malabsorption in Eurasia: a problem in culture history. *Anthropos* **74**, 61–80.
Tandon, R.K., Joshi, Y.K. Singh, D.S., Narendranthaw, M., Balakrishnan, V. and Lal, K. (1981). Lactose intolerance in North and South Indians. *American Journal of Clinical Nutrition* **34**, 943–946.
Wang, Y., Yan, Y., Xu, J., Du, R., Flatz, S.D., Kuhnan, W. and Flatz, G. (1984). Prevalence of primary adult lactose malabsorption in three populations of northern China. *Human Genetics* **67**, 103–106.
Weiss, K.M. (1996). In search of times past: gene flow and invasion in the generation of human diversity. In *Biological Aspects of Human Migration*, ed. C.G.N. Mascie-Taylor and W.G. Lasker, pp.130–166. Cambridge: Cambridge University Press.
Zvelebil, M. (1995). At the interface of archaeology, linguistics and genetics: Indo-European dispersals and the agricultural transition in Europe. *Journal of European Archaeology* **3**, 33–70.

# Index